NO

23. NOV 09.

17. DEC 09

30. 12. 09.

Please return/renew this item by the last date shown

worcestershire
countycouncil
Libraries & Learning

MEGADISASTERS

MEGADISASTERS

PREDICTING THE NEXT CATASTROPHE

FLORIN DIACU

OXFORD
UNIVERSITY PRESS

OXFORD
UNIVERSITY PRESS

Great Clarendon Street, Oxford OX2 6DP

Oxford University Press is a department of the University of Oxford.
It furthers the University's objective of excellence in research, scholarship,
and education by publishing worldwide in

Oxford New York

Auckland Cape Town Dar es Salaam Hong Kong Karachi
Kuala Lumpur Madrid Melbourne Mexico City Nairobi
New Delhi Shanghai Taipei Toronto

With offices in

Argentina Austria Brazil Chile Czech Republic France Greece
Guatemala Hungary Italy Japan Poland Portugal Singapore
South Korea Switzerland Thailand Turkey Ukraine Vietnam

Oxford is a registered trade mark of Oxford University Press
in the UK and in certain other countries

Published in the United States
by Oxford University Press Inc., New York

© Florin Diacu 2009

British Library Cataloguing in Publication Data

Data available

Library of Congress Cataloging in Publication Data

Data available

Library of Congress Control Number: 2009930120

Typeset by SPI Publisher Services, Pondicherry, India
Printed in Great Britain
on acid-free paper by
Clays Ltd, St Ives plc

ISBN 978-0-19-923778-4

1 3 5 7 9 10 8 6 4 2

I dedicate this book to all those who try
to make this world a safer place

You can only predict things after they have happened.
Eugène Ionesco

Predicting the future is easy. It's trying to figure out what's going on now that's hard.
Fritz R. S. Dressler

We have redefined the task of science to be the discovery of laws that will enable us to predict events up to the limits set by the uncertainty principle.
Stephen Hawking

CONTENTS

PROLOGUE:
GLIMPSING THE FUTURE

Prediction is very difficult, especially about the future.

Niels Bohr

A couple of years ago, a friend asked me whether I wanted to know my future. I was puzzled. The itch of curiosity enticed me to say yes, but the fear of bad news urged me to say no.

'I'd like to know only events I could affect.'

'You don't believe in fate then?'

'Outside factors lead our lives, no doubt, but our actions matter too.'

'Still, we never know what happens next.'

'Sometimes we can forecast things.'

'Yeah, right—like the weather in Victoria,' my friend laughed, hinting at the fact that meteorologists often make inaccurate predictions in our area.

Though it ended in disagreement, this discussion triggered an idea in my mind. What events can we predict? When are forecasts possible and how are they made? I thought of writing an informative and entertaining book aimed at readers with little or no science training, but who are willing to learn a few things about this topic.

My interest in prediction runs deeper. I am a mathematician. My research field is the theory of differential equations, which provides a language for the laws of nature. In particular, I work in celestial mechanics—a branch of mathematics and astronomy that tries to explain how stars, planets, and other cosmic objects wander in the universe. The motion of these bodies can be established by solving certain mathematics problems.

Celestial mechanics can predict the exact positions of all the planets thousands of years from now, forecast the day and time when eclipses take place, and detect invisible solar systems by studying the motion of stars. So I knew that it is within our power to predict celestial motions through careful reasoning and computations. But I also had motives to agree with my friend's concerns about predicting other phenomena.

My concern was with a property called *chaos*, which occurs in many dynamical systems. To mathematicians, chaos is another name for high instability: *similar starts don't guarantee similar outcomes.* Imagine, for instance, a trip on the Amazon with two rafts that float freely down the river. No matter how close to each other they start, the rafts drift, and the distance between them increases in time.

Examples of mathematical chaos abound around us. Leaving for work a few minutes later than usual can get people into the rush-hour traffic and significantly delay their arrival. Or, even though they share the same genes and upbringing, twins may live very different lives. In all these cases, no matter how close two evolving states begin, they may diverge from each other.

Therefore chaos makes predictions difficult. It may act fast, as it does with the weather, which cannot be forecast more than a few days in advance, or it may set in slowly, as happens with planetary motion, whose prediction becomes unreliable only millions of years later.

Studying all chaotic phenomena and finding out which of them allow reliable forecasts would have been a gigantic task. Therefore I wanted to focus on a few practical issues. So what should I opt for?

I decided to study phenomena that could affect the lives of many people. From here, the idea of researching megadisasters came naturally, and it was fairly easy to select the ones I would include in this book.

The television images of the 2004 Indian Ocean tsunami were still fresh in my mind, so I knew that these killer waves would be on my list of subjects. After all, the wave equation was part of a third-year course I taught at the University of Victoria. I had to dig into the history of the problem and find the connection between this differential equation and the work done on predicting tsunamis. In the company of pioneers of wave theory, like the mathematicians Lagrange and Laplace and the physicists Rayleigh and Fermi, this subject promised to be exciting.

Earthquakes formed an equally interesting topic. The undulation of the Earth's crust is also described by a wave equation, which meant that I was in my element again. Moreover, I had lived through earthquakes and had read about the attempts to predict them—an issue that is filled with controversy. Some scientists say forecasts can be made; others think the opposite. But in 1975, a strong earthquake was predicted in China about six hours before it happened. More than 150,000 lives were saved thanks to this warning. No doubt, something intriguing was going on here, and I had to find out what.

The memory of a trip to Italy, where I visited Pompeii, and later climbed Mount Etna in Sicily, played a role in including volcanic eruptions in my plans. But there were other reasons that influenced this decision. One of them was about the 1980 eruption of Mount St Helens, south of Seattle. The explosion had been heard in Victoria, which lies 300 kilometres north of the volcano. The timely forecast of this event saved many lives. Another incentive for wanting to research this subject was the famous Krakatoa eruption of 1883, which ejected twenty-five cubic kilometres of ash and rock and produced a tsunami that killed 36,000 people.

The 2005 Hurricane Katrina convinced me that cyclones, typhoons, and hurricanes should make my list too. From the mathematical point of view, these phenomena are studied in the framework of fluid mechanics, and I was well acquainted with the differential equations describing them.

The issue of climate change was a clear choice from the beginning not only because of the attention it receives today. Two colleagues of mine at the University of Victoria had expressed very different views on this topic. One was Andrew Weaver, an award-winning climatologist, the other—Jeff Foss, a philosopher. While Weaver, who runs his climate models on powerful computers, considers global warming imminent, Foss deems the dangers exaggerated. Since I know Andrew and Jeff, and respect them both, I decided that climate change was an appealing subject to consider.

Cosmic impacts are related to my expertise, and I wanted to approach this issue too. Several books discuss the problem of predicting such events, but they don't always agree on what should be done if a comet or an asteroid were to hit the Earth.

Therefore I had to understand which solutions were more reasonable, and perhaps suggest new ways of dealing with this problem. Moreover, I realized that governments don't take the cosmic threat seriously. Consequently, research done in this field is underfunded.

An issue I felt compelled to include in my study was that of stock-market crashes. In 1929, a sharp drop in stock prices marked the beginning of the biggest economic depression of all times and, combined with a shaky geopolitical situation, led to the most devastating war in human history. Hundreds of millions of people were affected worldwide. Yale economist Robert Shiller showed that conditions similar to those of the big depression occurred in 1999, and said that the world's economy was on the brink of a breakdown. A few days after he published a book on this subject, the market fell sharply, but luckily not low enough to produce a global catastrophe. Can we predict the likelihood of such events and take measures to avert them? At the time I was researching this issue, I didn't know that a disaster was threatening us. But as I went deeper into the problem, signs of potential trouble began to emerge.

As a mathematician with interest in the physical sciences, I wanted to know more about pandemics. Their prediction has less to do with medicine than with biology and mathematics. Indeed, mathematical biology is a field that has recently made remarkable progress. I was familiar with some of the technical models that help biologists in their research and wanted to see how the collaboration between mathematicians and epidemiologists could help prevent the spread of influenza or some other deadly disease.

The struggle to comprehend the issues mentioned here paid off. Now I know much more about predicting megadisasters than I knew when I started this project. I also feel privileged to have had the backing of two exceptional publishers. Princeton University Press took on the task of conveying my ideas to the North American public and Oxford University Press prepared the edition for the rest of the English-speaking world. And I am glad that a top Japanese publisher supported this project long before it was finished.

That's how this book was born. Let us now follow its windings in the quest for a safer planet.

1. WALLS OF WATER: TSUNAMIS

I got outside my hotel, and saw that the ocean was now level with our island. To my horror, a wall of water—boiling, frothing, angry as hell—was bearing straight down at us, and a strange mist that looked like thick fog blocked out the sun. I stopped breathing...

*Dave Lowe, eyewitness of the 26 December 2004
tsunami on the South Ari Atoll in the Maldives*

We relate Christmas to happiness, but no holiday can shield us from grief. On the night of 25 December 2004, some breaking news shook North America. A catastrophe had killed thousands of people in Southeast Asia, many foreign tourists among the dead. The number of reported victims was growing by the hour.

The rim of the Indian Ocean had been hit by a *tsunami*—also known as *tidal wave*—a tremendous shift of water that acts like a deluge. Waves of such force are triggered by marine earthquakes, landslides, and volcanic eruptions, or by large meteoritic impacts. While in deep waters, tsunamis might pass undetected because of their long and gentle shape. But once the seabed shallows, they swell and invade the shore with a force that may flatten the ground.

I will never forget the images shown on television: the incoming wave, the water rushing through the windows of a restaurant, the old man swept away from the terrace of his hotel, the woman trying to cling to the branch of a palm tree, the father and the child running for their lives, the scream of the desperate mother, the indigenous boy rescuing a blonde girl from the flood...

There were many stories, most of which I have forgotten— stories of loss, grief, hope, or happy reunion. But one of them, which I heard months later, stayed with me. It was the tale of a survivor, a story told with inner peace and resignation during a Larry King Live show on CNN. This is what I learned from it.

The Model and the Photographer

Petra Nemcova and Simon Atlee spent their Christmas holiday in Khao Lak, a lavish beach resort in southern Thailand. Petra was a Czech supermodel, and Simon a British photographer. They had fallen in love while he was shooting pictures of her for a fashion magazine. But because they travelled on different assignments, they hadn't seen much of each other during the past few months.

This vacation had been Petra's idea. She found Thailand amazing—a country with wonderful people, a soothing climate, and breathtaking landscapes. The trip was meant to be a surprise for Simon, so she told him about it only shortly before their departure.

Christmas Day went by peacefully. They tanned on the beach and talked about marriage and children. The wedding date was something they had still to set. After dinner they went to their room to watch *White Christmas*, the 1950s musical comedy with Bing Crosby, Danny Kay, Rosemary Clooney, and Vera Ellen.

Petra had not seen the movie before, and Simon thought she would like it.

The next morning they woke up early. Their stay at this orchid resort had come to an end, and they wanted to get ready for departure. But first they had breakfast and took a walk along the beach. On return, Petra started packing. Simon went for a shower. Then tragedy hit with almost no warning.

Through the balcony window Petra saw people running away from the beach. They were screaming in panic as if a noisy marine monster were following them.

'What's happening?!' Simon shouted from the bathroom.

'I don't know! An earthquake or something!'

Seconds later the glass window broke. In no time, the tsunami blew up their bungalow and swept them away.

'Petra!! Petra!!' Simon cried.

'Catch the roof!' Petra called out before she was pulled under a swirl of dirty water.

Debris hit her, tore off her clothes, and she felt a strong pain in her pelvis. When she resurfaced, Simon was nowhere to be seen. Then the wave covered her again.

She thought she would die. Hope revived when she came close to a palm tree, but her attempts to cling to it failed. Luckily another tree appeared in her way, and with great effort she grabbed one of its branches. Although debris hit her repeatedly, assailing her naked, battered body, she clung to the trunk. Desperate voices could be heard from neighbouring trees.

As the first shock receded, Petra thought of Simon. He was a good swimmer, so she hoped that he had made it to a safe spot.

She prayed for him, and she prayed that the tree holding her would stand the force of the stream.

Time passed. Petra often had the illusion that this was just a nightmare from which she would awake soon, but the pain brought her back to reality. Although she felt very tired and her arms had grown numb, she knew that she had to stay put. Between ocean and sky, her life hung in the balance.

Eight hours later, two courageous Thai men rescued her. They had to handle her carefully because every move made her cry. She would go through a lot of pain in the days to come. Fortunately the immediate danger had passed. She spent several weeks in a Thai hospital with internal injuries and a shattered pelvis, and needed several months to recover completely.

But Petra never saw Simon again. Some human remains found in March 2005 were identified as his. He met the fate of the more than 200,000 people who happened to be in the path of destruction on that godforsaken day. The saddest part of the story is that most of those lives could have been saved.

How It Happened

On 26 December, at 6.58 a.m. local time, an earthquake shook the Indian Ocean, off the Indonesian coast of Northern Sumatra, 250 kilometres southeast of Banda Aceh. Initial estimates put its magnitude at 9.0. The shock was felt as far as the Bay of Bengal. The earthquake occurred between the India and Burma plates as the former shifted beneath the latter, raising the ocean's bottom by ten metres in some places. This event triggered a tsunami, which hit the beaches bordering the Indian Ocean in

Indonesia, Sri Lanka, India, Thailand, Somalia, Myanmar, the Maldives, Malaysia, Tanzania, Seychelles, Kenya, and Bangladesh (Figure 1.1). No tsunami ever has claimed so many lives.

Some scientists flew to Indonesia to learn more about the cause of the disaster. Others began to analyse the data. Richard Gross, a geophysicist with NASA's Jet Propulsion Laboratory, reported that a shift of mass towards Earth's centre had caused the planet to move one millionth of a second faster and tilted its axis at the poles by an inch. Seismologists Seth Stein and Emile Okal of Northwestern University claimed later that the earthquake had been much larger than initially thought, namely 9.3 on the *moment-magnitude scale*, for which a one-point increase corresponds to about a thirty-fold effect.

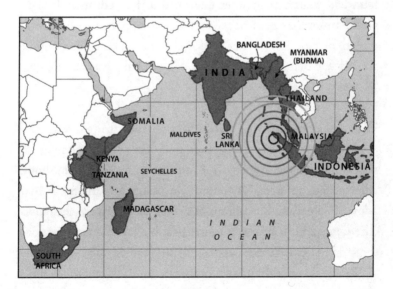

Figure 1.1 The shores affected by the Indian Ocean tsunami on 26 December 2004.

Such re-evaluations are not unusual. The rupture zone had been bigger than reported, the initial estimates ignoring the slower shifts along the fault. To extract these data, Stein and Okal relied on theoretical results they had developed three decades earlier with Robert Geller, now a professor at the University of Tokyo.

Shortly after the earthquake, Sumatra's coast was hit by a wall of water higher than the coconut palms lining its beaches; the tsunami, however, travelled almost two hours before reaching Thailand, India, and Sri Lanka. A warning procedure, like the ones used in North America and Japan, might have reduced the casualties to a minimum. Alas, such a system was non-existent in the affected zones.

The ideal scenario would have been to forecast the tsunami and take suitable measures days or hours in advance. But are such predictions possible?

Solitary Waves

To forecast events, we must know how they form and develop and what laws govern them. Tsunamis occur rarely and look like big wind-generated waves, but instead of breaking at the shore, they go inland. Progress towards understanding them has been slow. The nature of tsunamis remained unclear until the end of the nineteenth century. All their possible causes became apparent only several decades ago.

Research on solitary waves began in August 1834 when a young engineer named John Scott Russell conducted some experiments on the Union Canal near Edinburgh in Scotland.

The railroad competition threatened the horse-drawn boat business, and Russell had to assess the efficiency of the conversion from horsepower to steam. In his report, he described the following occurrence.

As a rope got entangled in the device used for measurements, the boat suddenly stopped and the water 'accumulated round the prow of the vessel in a state of violent agitation, then rolled forward with great velocity, assuming the form of a large solitary elevation—a rounded, smooth and well-defined heap of water—which continued its course along the channel without change of form or diminution of speed'.

This *wave of translation*—as he called it—intrigued him, so he 'followed it on horseback, and overtook it still rolling on at a rate of eight or nine miles an hour, preserving its original figure some thirty feet long and a foot to a foot and a half in height' until he lost it in the meanders of the channel. This event was the start of a struggle to understand an unusual phenomenon and—what would be an even more difficult task—to prove the existence of water waves that could travel forever.

In 1830 he invented a steam carriage, but his undertaking failed because the officials opposed its implementation. Russell had more success with the Union Canal Company, which hired him to study the connection between wave generation and resistance to motion. This opportunity had also been triggered by chance. When a horse dragging a boat on a Glasgow canal took fright and ran off, the vessel's prow rose and the boat sailed faster. Russell understood that the solitary wave caused the reduced resistance and the rise of the boat, so he focused his research on the wave.

He built a water tank, generated waves of translation by releasing a column of water through a sliding panel, and performed hundreds of experiments, recording the details he observed. Although the wave's fast speed was remarkable, Russell was more impressed by its persistence. He had expected the wave to shrink after travelling long enough, but the tests proved him wrong. The solitary wave looked more stable than anything he had seen before.

The wave of translation appeared only if the boat reached a critical speed. Below it, the vessel met water resistance; above it, the wave became self-sustained, allowing the boat to move easier. After repeated experiments, Russell concluded that the wave's velocity depends both on the depth of the water and on the wave's height.

His result explains why tsunamis move at high speed in mid-ocean but slow down close to the shore and why boats overcome water resistance in shallow canals as they reach the critical speed. In deep seas, however, ships are slow, moving well below the critical speed, so by trying to move faster they encounter more resistance. Russell solved this problem by designing hollow-lined prows, which part the water without ruffling its surface. He noted that pirates, to whom speed was essential, had built similar prows in the past.

Russell also studied the interaction between waves. Intuition suggests that, at impact, waves travelling in opposite directions break. But this never happens. They meet, merge for an instant, and pass through each other unchanged. This phenomenon shows why the idea to kill a tsunami through a collision with an artificially generated wave doesn't work.

Apart from conducting some 20,000 experiments with toy models and ships ranging from a few hundred grams to

1,300 tons, Russell spent years analysing the shape and motion of the translation wave. Among other things, he learned that, unlike wind-generated waves, which involve vertical motion, the solitary wave is a horizontal shift of mass with a shape about six times longer than tall. So instead of moving up and down, as ordinary waves do, a tsunami pushes ahead like a shelf of water.

Russell also had an original idea about tides, which he viewed as very large solitary waves. He divided his tidal theory in two parts, one founded on celestial mechanics, to explain water elevation in seas and oceans, and the other based on hydro-dynamics, to account for the swell of small basins, rivers, and canals.

Russell presented his research in several articles, which he submitted to different meetings attended by mathematicians, physicists, engineers, and astronomers interested in fluid dy-namics. Among them was George Biddell Airy, who opposed Russell's results. Airy had a theory of his own, and he deemed the solitary wave impossible.

Meeting Resistance

No fancy idea permeates the scientific world with ease, particu-larly so when a personality opposes it. Airy was no ordinary scientist. He held the Lucasian professorship at Cambridge, a position Isaac Newton had occupied in the seventeenth century, and had the most envied astronomical job in Britain, that of Astronomer Royal. He would later preside over the Royal Society and accept a knighthood, but only after declining it three times because he could not afford the fees.

Airy made important contributions to science, from improving the orbital theory of Venus and the Moon to a mathematical study of the rainbow. He preferred applications to theory, and was often at odds with his colleagues about the research direction in which important mathematics prizes should go.

Outside his professional work, Airy showed broad interests. He read history and poetry and was keen about architecture, geology, engineering, and religion. He even tried to identify the location of Julius Caesar's landing in Britain and the place from where the Roman consul departed. But later in life he spent most of his time in administration.

When Russell announced his results, Airy was still very active in research. The Astronomer Royal had constructed his own theory of waves, which he had initially based on the work of the French mathematician Pierre Simon Laplace. But because Laplace's equations applied only to shallow waves, Airy came up with some improvements. His goal was to predict the height of tides. Alas, he failed in this endeavour as much as his French predecessor did; their calculations didn't come even close to reality.

Airy, however, considered his theory suitable for understanding waves. In an article published in 1845, he praised Russell's experiments because his own theory explained them. But he warned against Russell's analysis. The equations Airy developed could not account for large shifts of mass, so an everlasting wave made no sense to him.

Although this authoritative judgment failed to shake Russell's belief in the value of his discovery, the Scottish engineer received another blow soon. In 1846, the new leader of British hydrodynamics, Cambridge mathematician George Gabriel Stokes, published a paper about the state of the field. Stokes's point was clear: permanent translation waves could not exist.

Further Opposition

Aged 27 at that time, Stokes was eleven years younger than Russell, and his paper confirmed his leading role in the field. In 1846, when this report appeared, he could not accept the idea of a sea wave that travels thousands of miles undisturbed. In his opinion, waves of translation had to shrink, and the stability Russell proclaimed was illusory because he had drawn his conclusions from experiments performed in short tanks.

Stokes's interest in waves faded soon, but he returned to them time and again. In October 1879 he wrote to William Thomson, better known as Lord Kelvin: 'I have in mind when I have occasion to go to London to take a run down to Brighton if a rough sea should be telegraphed, that I may study the forms of waves about to break. I have a sort of imperfect memory that swells breaking on a sandy beach became at one phase very approximately wedge-shapes.' When Kelvin invited him 'to see and *feel* the waves' on his yacht, Stokes answered in September 1880:

You ask if I have done anything more about the greatest possible wave. I cannot say that I have, at least anything to mention mathematically. For it is not a very mathematical process taking off my shoes and stockings, tucking up my trousers as high as I could, and wading out into the sea to get in line with the crest of some small waves that were breaking on a sandy beach.

Stokes's change of mind about the value of practical observations stemmed from his new conviction that Russell had been right. Three weeks later, he wrote to his friend again: 'Contrary to an opinion expressed in my report [of 1846], I am now disposed to think there is such a thing as a solitary wave that can be theoretically propagated without degradation.' Kelvin disagreed. His opposition resided in some technicalities related to the mathematical model, which was not transparent to the subtleties of solitary waves.

Russell never learned about this exchange or of Stokes's approval of his work. At that time, he had lost his ambitions. In the 1860s he had suffered several blows: his attempts to build an iron vessel called *Great Eastern* failed, he got involved in a financial dispute about an armament contract, and was expelled from the Council of the Institute of Civil Engineers. These setbacks made him withdraw on the Isle of Wight, in southern England, where he died on 8 June 1882, aged 74.

At the time of these developments, unknown to Russell, Kelvin, and Stokes, a young French mathematician named Joseph Boussinesq was also studying solitary waves.

The French Connection

Two mathematical giants—a Frenchman, Jean Le Rond d'Alembert, and a Swiss, Leonhard Euler—had initiated the study of waves more than a century earlier. In 1747, d'Alembert won the prize of the Prussian Academy for his pioneering work on partial differential equations, which model many physical phenomena, including waves. In spite of its sound mathematics, however, the paper's physics left much to be desired. For instance, d'Alembert erroneously stated that tides, not heat, generate winds.

Nevertheless, Euler saw the value of d'Alembert's methods, developed them further, and found their true physical meaning. But the Swiss mathematician gave d'Alembert little credit, a fact that triggered animosity between them. In spite of this rivalry, both d'Alembert and Euler followed each other's papers to lay the foundations of wave theory. Their ideas were further developed by two other mathematical geniuses: Joseph Louis Lagrange and Pierre Simon Laplace.

In 1776 Laplace published his celebrated theory of tides, which was based on a model for the propagation of small waves in shallow water. Lagrange extended Laplace's ideas to deep-water waves, but many of his assertions were speculative. The two mathematicians showed more interest in the shape of the surface than in how the fluid moved. Consequently, neither of them could guess the existence of the solitary wave.

A step forward was made in 1813, when the Academy of Sciences established a prestigious prize. Laplace drafted the question: 'An infinitely deep fluid mass, initially at rest, is set into motion by a given force. It is asked to determine the form of the external surface of the fluid and the velocity of every molecule on the surface after a given time.' The mathematician who came close to solving the problem was Siméon Denis Poisson, the most brilliant disciple of Laplace. In his mid thirties at that time, Poisson was already established as a professor at École Polytechnique and astronomer at Bureau des Longitudes in Paris. But since he was also a member of the jury, Poisson could not compete for the prize.

The award went in 1816 to the 27-year-old Augustin Louis Cauchy, best known today for putting calculus on a rigorous foundation. Notable is the significant overlap between the results of Poisson and Cauchy, though they worked independently. Both men used a theory their countryman Jean Joseph Fourier had developed a decade earlier, but unlike Poisson, Cauchy recreated most of its tools from scratch.

A problem Cauchy analysed was the disturbance created by the sudden immersion of a solid body into a fluid. This issue was close to the study of a boat's motion on water. Cauchy, however, didn't go far enough to see any connection with the solitary wave. His failure comes as no surprise: nobody before Russell, and very few of Russell's peers, thought that this phenomenon existed. Among those who did was the French mathematician Joseph Boussinesq, who succeeded in going beyond Russell's practical experiments, laying the foundations of a solid theoretical framework.

In 1871, four years after obtaining his doctoral degree, Boussinesq became acquainted with Russell's work and the wave experiments of the French hydraulician Henri Bazin. Boussinesq had already rediscovered some known results. This exercise now led him to the solitary wave, whose existence he could prove. His solution was based on an ingenious method applied to Euler's equations. Although Lagrange had used the same idea a century earlier, he failed to push it through because he lacked the more refined mathematical techniques Boussinesq employed.

But Boussinesq was not the only one who rigorously proved the existence of the solitary wave. Five years later, independently of him, an English mathematician reached the same conclusion.

His name was John William Strutt or Lord Rayleigh. Russell, Stokes, and Kelvin knew nothing of this feat.

The Tricks of History

John William Strutt was born the same year as Boussinesq, 1842, in Witham, Essex, as the son of the second Baron Rayleigh of Terling Place. Nobody in his family of landowners showed any interest in science, and John proved to be average in school. But his talent for mathematics sprouted in 1861 after he began to study at Cambridge. Stokes influenced his intellectual development, though he never encouraged Strutt to pursue research. After graduation, Strutt was elected a Fellow of Trinity College, and at the age of 30 inherited his father's title.

In spite of travelling widely as a young man, Rayleigh dedicated much time to research. His theory of scattering, published in 1871, gave the first correct explanation of why the sky is blue. In 1879 he was appointed Cavendish Professor of Experimental Physics at Cambridge. This opportunity helped Rayleigh lay the foundations for his work on the density of gases and put him on the path of discovering the chemical element argon, contributions for which he was awarded the Nobel Prize for Physics in 1904.

While in his early thirties, Rayleigh became interested in wave theory and in Russell's work. He started like Boussinesq from Lagrange's principles, but then distanced himself from the French school. Rayleigh had an idea that simplified his computations. Unlike Boussinesq, he looked at the wave as if riding on it, using a coordinate system bound to the wave, not fixed somewhere in space.

Some twenty years later, in 1895, the Dutch mathematician Diederik Johannes Korteweg and his doctoral student Gustav de Vries improved Rayleigh's method and extended it to the study of other types of waves, including the oscillatory ones and those of evolving shape. The equation they derived was a version of Boussinesq's.

Korteweg and de Vries also found the solution that corresponds to the wave of translation, and duly mentioned the priority of Boussinesq and Rayleigh. But they showed no excitement about this rediscovery, seeing little merit in going where others had been. Like most mathematicians, they were more interested in obtaining original results, and cared less about what applications those would have. Thus they emphasized 'the new type of long stationary waves', called *cnoidal* today, which was their discovery.

The paper of the Dutch mathematicians made no immediate impression. Their peers considered the problem of the solitary wave completely solved, and were seeking new (and more exotic) phenomena related to the motion of liquids. Later, when the experts understood the significance of this equation, they called it *Korteweg–de Vries*.

History played a trick on Boussinesq, who is only briefly mentioned in the field. For the two Dutch mathematicians, this was their most important achievement. Their merit cannot be denied, but they didn't excel in research as much as Boussinesq and Rayleigh did. Korteweg had become a professor at the University of Amsterdam in 1881, from where he retired in 1918 without publishing other important papers.

De Vries achieved even less. He gave up research after writing two more articles about cyclones and made a living as a schoolteacher.

The Korteweg–de Vries equation, however, took on a life of its own. If, during the first decades of the twentieth century, it didn't stand out in any particular way, this was about to change. An apparently unrelated experiment would propel it to the limelight of mathematical research. The reason for this turn of events was the role played by an Italian physicist, also known as 'the father of the atomic bomb'.

A Numerical Experiment

Enrico Fermi, the key nuclear physicist of the Manhattan Project, also had a crucial contribution to the theory of waves. It all started a few years after the war, during one of Fermi's visits to Los Alamos. Among the friends the Italian scientist had made while working on the bomb was the Polish mathematician Stanislaw Ulam. Fermi had used his 1938 Nobel Prize award ceremony in Stockholm to flee Fascist Italy and cross the Atlantic Ocean. Ulam had come to the United States a few months before Fermi as a Harvard Junior Fellow. When his contract expired, the Pole received a faculty position at the University of Wisconsin, where he engaged in mathematical research with military applications.

In the middle of the War, Ulam received an invitation to join a secret project in New Mexico. He was told little about it, so he felt no wish to go. But after learning the names of those who had recently vanished from campus, and suspecting a connection with the project, he agreed to go. This decision changed his life, propelling him into the circles of the world's mathematical elite. Part of his rise to fame was due to his collaboration with Fermi.

The post-war American scientific establishment was split about the newly invented electronic computers. Fermi took the side of those who welcomed them. In the early 1950s, he began to think about finding some significant physics problem that would merit an investigation on one of the very first electronic computers, MANIAC I, which happened to be located at Los Alamos. He decided to study how crystals evolve towards thermal equilibrium.

Crystals are modelled as particles kept together by forces. The ensuing structure resembles the builder's scaffolding—the bars corresponding to the forces and the junctions to the particles. When not in equilibrium, crystals behave like structures whose bars bend and stretch. This three-dimensional model reflects reality quite well. A simpler, less realistic model is a two-dimensional one, which would look more like a chicken wire or a fishing net. A one-dimensional model would resemble a chain.

Apparently there is no overlap between waves and crystals, and Fermi never thought to connect them. But a bit of imagination reveals that a wave's surface looks like a bed sheet, which is nothing but a fishing net with tiny holes. Similarly, the spatial waves resemble 'soft' scaffoldings with many joints and short bars. Therefore understanding the motion of crystals could shed light on the propagation of waves.

To keep computations easy, Fermi and Ulam decided to work on the one-dimensional case—the chain. However, in order to implement a suitable numerical technique, they needed an expert in computational methods. Thus, a third Los Alamos expert joined the team. His name was John Pasta.

Fermi, Pasta, and Ulam drafted a research plan in 1952 and did their first numerical experiments the following summer. It was Fermi's idea that, instead of performing the standard calculus for a physical problem, one should take particular examples and test them with computer simulations. This approach would not only stimulate the development of wave theory but also prove crucial in the study of non-linear phenomena—one of the main research directions in the mathematical sciences during the last three decades of the twentieth century.

The numerical experiments on the 'chain crystal' revealed the existence of a solution that looked like the solitary wave, as if a snake curved its body into a bump that moved smoothly from head to tail. Fermi was not impressed, but he changed his mind soon. This revelation was probably his last. During the second half of 1954, the stomach cancer he was suffering from began to spread, and he died at the end of November, less than two months after turning 54.

Solitons

In 1965 two engineers, Norman Zabusky of Bell Telephone Laboratories and Martin Kruskal of Princeton University, showed the Fermi–Pasta–Ulam experiment and the Korteweg–de Vries equation to be different sides of the same coin. They also brought numerical evidence that the solutions of the Korteweg–de Vries equation obeyed a property that Russell had observed in experiments more than a century earlier: translation waves pass through each other undisturbed.

It took a few years until the results of Zabusky and Kruskal were recognized, but starting with the early 1970s, their paper had a strong impact on wave theorists, many of whom put their current research aside and jumped to the study of the Korteweg–de Vries equation, which had remained obscure for several decades. The experts felt there was something new to discover. They reopened a research direction that had made little advance during the first half of the twentieth century.

This was not the first time a mathematical field saw an unexpected revival. Chaos theory had a similar fate. At the end of the nineteenth century, the French mathematician Henri Poincaré pointed out the existence of deterministic phenomena whose high instability makes their long-term evolution difficult to predict. Very few people understood him, so this part of his research was neglected. The Fermi–Pasta–Ulam experiment influenced its resurrection.

Edward Lorenz, a meteorologist at the Massachusetts Institute of Technology, was one of the key players in the revival of chaos theory. In 1972 he addressed the American Association for the Advancement of Science with a talk entitled 'Predictability: Does the Flap of a Butterfly's Wings in Brazil Set Off a Tornado in Texas?' He explained why instability might amplify a small cause into a large effect. Thus, the butterfly metaphor and the word *chaos* aroused the interest of many mathematicians.

In their paper, Zabusky and Kruskal coined a magical word too, the *soliton*—short for the 'solitary-wave solution' of the Korteweg–de Vries equation. As it happened to chaos theory, the combination of an inspired word (soliton) and a surprising property (wave-crossing) triggered a research boom in the study of waves. Zabusky and Kruskal were the tipping point in the development of the field.

But once this new branch of wave theory sprouted, the mathematicians focused on understanding the abstract world of the Korteweg–de Vries and other related equations, neglecting the solitary wave. Those who didn't lose the original track were the engineers. They followed the latest mathematical results and applied them to tsunamis.

From Theory to Practice

Russell and his immediate followers never saw a real tsunami. Neither is there any indication that they heard about the formation of translation waves in the world's oceans. Those who witnessed such events were unaware of the scientists' efforts, so the bridge between theory and practice was a slow one to build. Moreover, tsunamis are more complicated than solitons, which resemble them only in a first approximation. But the contributions mentioned earlier laid the foundation for understanding these waves.

A first difficulty engineers encountered was the non-linearity of the Korteweg–de Vries equation. Although Henri Poincaré had initiated the study of non-linear phenomena at the end of the nineteenth century, the field boomed only in the 1960s, triggered by contributions such as those of Lorenz, Fermi, Pasta, and Ulam. Thus the first attempts to grasp the generation and propagation of tsunamis were restricted to linear models, which approximated Korteweg–de Vries-type equations.

Work started in the 1950s in Japan, a country often hit by tsunamis. (In fact the word 'tsunami' is Japanese for 'harbour wave', and was used by the fishermen who returned home to devastated harbours.) The early studies determined the general wave pattern near the source region for a variety of bed motions in simplified situations. These attempts continued in the 1960s for more realistic models, which allowed good estimates for the size of tsunamis relative to how much the seabed changes during an earthquake.

The transition from linear to non-linear models took place gradually. Various authors employed linear equations with some non-linear perturbation terms. Their results did not add much knowledge about tsunamis, but led to a refining of the mathematical techniques used until then. These achievements, together with the independent development of non-linear analysis in the early 1970s, allowed experts to study the formation and propagation of tsunamis in the more general and realistic framework of non-linear theory—the research direction followed today.

Physics also plays an important role in understanding tsunamis. For instance, the recent analysis done by Jack Hills and Patrick Goda of the Los Alamos National Laboratory in New Mexico showed that the marine impact of a 500-metre meteorite could produce a one- or two-kilometre high tsunami, which might penetrate hundreds of kilometres inland. Critics, however, are sceptical that their model matches reality and think that waves this high are impossible on Earth. Independently of height, the probability of such events is small, but—as Chapter 6 will show—far from zero.

Other directions of research have identified dangerous geographic zones. When such places are populated or economically important, the communities can prevent potential disasters by building anti-tsunami walls, as was done in Japan, or by planting forests on the shore to dissipate the wave's destructive energy.

During the past few decades, experts have also learned about the circumstances triggering tsunamis. If the cause is an earthquake, for example, its magnitude must be significant to create a deadly wave. But when the earthquake provokes landslides that generate tsunamis, then even lower magnitudes are sufficient to produce large waves.

None of these scientific efforts, however, can tell precisely when a tsunami will strike. The reason for this drawback stays with the triggering event, which precedes the tsunami. Future chapters will discuss two of the generating factors: earthquakes and meteorite impacts, and describe the state of the art of their prediction.

The theoretical studies described here allow scientists to compute the size of a wave relative to the geography of the location, such that housing developments can be planned sufficiently high above the sea level. Combined with observations and experiments, these achievements have led to the discovery of the signs that warn about an incoming tsunami at least a few minutes in advance. Moreover, North America and Japan implemented warning systems, which allow the quick evacuation of the potentially affected zones whenever a triggering event happens. Practice shows that if there is enough time left, the evacuation can be effective.

Educating people about the signals indicating the approach of tsunamis can save lives. We don't have to look too far to understand the advantage of recognizing the danger and acting decisively. The recent Indian Ocean disaster offers a stunning example.

The English Girl

On the morning of 26 December 2004, Tilly Smith was with her parents and younger sister on the Mai Khao beach in Phuket, not far from where Petra Nemcova and Simon Atlee were getting ready for departure. Ten years old, blond hair falling on her shoulders, Tilly looked jolly and fresh as she smiled in the sun. She had good reasons to be happy, thinking of the damp, cold England she had left behind.

As mother and daughter went down to the water, Tilly noticed that the sea looked bubbly and frothy like on the top of a beer. This image triggered some recent memories in her mind. Two weeks earlier, during her geography class at Danes Hill School in Oxshott, Surrey, Tilly saw a movie about the Hawaii tsunami of 1946. She recognized the same warning signs: just minutes before the deadly wave hit, the water had begun to form bubbles and turn foamy, so very much as it did now.

'A tsunami is coming!' Tilly shouted. 'We must run away!'

Her mother, Penny, had noticed that strange phenomenon too, but thought it was due to a bad day at sea, so she didn't take Tilly seriously.

'Mummy, we must get off the beach *now!*' Tilly insisted.

Penny began to worry about her daughter, who started shouting madly at her, frustrated that her mother was blind to the danger. But in spite of her anxiety and irritation, Tilly didn't give up. She had to convince her mother somehow.

'If you're not coming,' she screamed, 'I'm going to leave you here!'

Penny remembered Tilly's recent geography class about tsunamis, so she gave in, and they both ran to Colin, Tilly's father. After he heard the story, Colin alerted a lifeguard, who evacuated the tourists from the beach. It was a wise decision. Minutes later the killer wave showed up.

In September 2005, back in England, Tilly received the Thomas Gray Award of the Marine Society in recognition of her timely and decisive actions, which saved human lives. But she most cherished the inner satisfaction that she and her family, as well as the about 100 people who had been on the Mai Khao beach, were alive because she alone had recognized the danger.

Warning Signs

Nothing happens without warning; it's just that the signs are often obscure or opaque. Heart attacks, for instance, take many by surprise because they do not recognize the symptoms. The same problem occurs with most natural disasters. But the difficulty to acknowledge the message and act on the spot can be eliminated through training.

Most tsunamis give notice tens of minutes before they reach the shore. One or several of the following signs may be seen, heard, or felt:

- an earthquake occurs
- the sea recedes to a considerable distance
- the sea bubbles
- the water stings the skin
- the sea smells of rotten eggs or oil
- a flash of red light sparks on the sea near the horizon
- a boom is followed by a whistle, or by a jet-plane- or helicopter-like noise

Each phenomenon depends on circumstances, and their explanations vary. The sea's recession compensates for the soon-to-arrive wave. The bubbling and the stinging of the skin, which resemble the effects of sparkling water, are consequences of the air or gases the wave pushes ahead of it. An earthquake or a volcano eruption may release chemical components with funny smells or lead to reactions that produce electrical discharges. Finally, the noises can be consequences of those reactions or may occur because of the friction between the large wave and the shallow seabed.

Some early warnings, which often occur before any of the physical signs are apparent, may come from nearby animals. The first documented animal reaction to a tsunami appears for the underwater earthquake of 1755, whose epicentre was some 350 kilometres southwest of Lisbon. The combined disaster killed between 60,000 and 100,000 people in Portugal, Spain, and North Africa, with the tsunami playing a lesser role in those deaths.

The earthquake struck on the morning of 1 November, causing wide fissures in Lisbon's centre. Those who rushed to the docks for safety watched as the receding water revealed a seafloor littered by shipwrecks and cargo. More than an hour after the earthquake, a twenty-metre high tsunami, followed by two more waves, engulfed Lisbon's harbour and downtown area. The animals are said to have run away from the shore long before the wave showed up.

The reason why animals feel the approach of tsunamis is yet unknown. Some scientists speculate that animals can detect certain earthquake waves, which propagate through the Earth's crust, as we will explain in the next chapter.

Nevertheless, warning signs exist for tsunamis. But we need to learn how to read them. This is not always easy because we may see danger when there is none. Aesop's tale about a shepherd boy who cried wolf to make fun of his fellow villagers shows how false alarms could do more harm than no warning at all. Nevertheless, Tilly Smith's example proves that instruction and vigilance can save lives.

So are there ways to forecast the events that may trigger tsunamis? After all, a big earthquake or a large meteoritic impact might be worse than a solitary wave. Their effect may range from killing half a million people to wiping out most life on Earth. The prediction of these phenomena is therefore worth pursuing.

2 LAND IN UPHEAVAL: EARTHQUAKES

If a seismologist fails to predict an earthquake, is it his fault?

Edwin Kobe

I experienced my first earthquake at the age of seven. I was playing alone in the living room when our building shivered slightly and a picture of my mother swung on the wall for a few seconds. I ran to the kitchen, where my father was doing some housework, and told him that I had just felt an earthquake. He smiled, petted me, and said:

'It must have been a heavy truck.'
'No, it was an earthquake! I saw Mummy's picture swinging!'
'Don't worry. Whatever it was, it's over now.'

Like most people in our city, my father didn't feel the shock, but the evening news proved me right. Since then, I have had several seismic experiences, and I have always recognized them the instant they started.

The strongest earthquake I have felt so far happened on the night of 4 March 1977. It had a magnitude of 7.2 and a depth of about ninety kilometres beneath the bend of the Carpathian mountain arc. I was at home in Sibiu, an old Transylvanian city, studying for a high-school exam. My parents were in the living room, chatting quietly so as not to disturb me. When the glass doors of my bookcase began to vibrate, I jumped to my feet and rushed to my parents.

'It's an earthquake!' I said. 'Let's get out of here!'

We lived on the ground floor of a downtown mansion with clay shingles, massive brick walls, and sturdy doorframes typical of Europe's medieval cities. A huge wooden gate guarded the large patio, which led to several apartments on each of the mansion's two levels. The rooms had high ceilings and tall terracotta stoves heated with natural gas.

My parents needed a couple of seconds to understand what was happening. My mother stood up, but my father sat calmly, as if weighing his options. As a veteran of the Second World War, he had been through much worse.

'Don't go outside!' my mother shouted. 'Shingles will fall from the roof! Let's get under the main doorframe!'

We followed her advice. Our neighbours had gathered under their doorframes too. Everybody but my father seemed scared. The mansion was shaking, and the patio was waving like a flying carpet. The tremor lasted almost a minute. Apart from a few cracked walls and fallen chimneys, nothing bad happened in our city, but thousands of people died in the south of the country, most of them in Bucharest. Many of those who escaped from the rubble became crippled for life.

For the first time, nature had made me feel helpless. I was almost eighteen, and that event awoke in me the curiosity to understand how earthquakes occur and whether we could ever predict them.

Between Hope and Confusion

On 4 February 1975 an earthquake of magnitude 7.3 struck the city of Haicheng in northeast China, killing more than 2,000 people, injuring almost 28,000, and damaging 90 per cent of the buildings. But in spite of this tragic outcome, the local authorities claimed victory. Six hours before the event, seismologists had warned about the disaster, and most inhabitants stayed outdoors. Had this prediction not been made, some 150,000 people would have died during the cataclysm.

The news pleased many people in Chile, Alaska, Japan, Tibet, and other areas of the world touched by major earthquakes in the recent past. But their hopes for a safer future were short-lived. On the night of 28 July 1976, a magnitude 7.6 earthquake struck the city of Tangshan, east of Beijing, apparently without warning. How many people perished is not known. Some sources put the number at 250,000, others indicate more than twice as many. In terms of casualties, this natural disaster has been surpassed only by another China earthquake, which apparently killed 830,000 people in 1556.

Contradictory news leads to confusion, but even probabilistic predictions can get people mixed up. On 2 February 2007, for instance, scientists warned of a higher quake risk in the region where now I live. A team of seismologists at the Geological Survey of Canada had noticed a series of slight tremors along Vancouver Island. This phenomenon was not new to them. They had registered it every fourteen months in the past few years. During the week when these tremors take place, major earthquakes are more likely, so the experts warned of a possible major shock in the days to come.

Garry Rogers, one of the scientists on the team, pointed out that there is no reason to panic. He came up with a metaphor to suggest the level of danger. We drive cars all the time knowing we could have an accident. We also accept that collisions are more likely during intense traffic. 'Well,' Rogers said, 'right now, we are in the middle of rush hour.' A front-page article published in Victoria's *Times Colonist* advised people to pack an emergency kit just in case something happens.

Two days later, however, the tremors stopped. It was as if we returned to normal traffic faster than usual. Of course, this fact did not exclude the possibility of an earthquake anytime, but it made it less likely until the next week of tremors, expected in April 2008. Can we relax? Yes. Drop our guard? No. Vancouver Island is close to the junction of two tectonic plates, where a magnitude 9 earthquake or higher happens, on average, every 500 years. The last one took place in January 1700.

'It looks like we dodged the Big One again,' started an editorial in Victoria's local newspaper on 6 February 2007. Another publication suggested that we were safe for another two centuries. I smiled when I read these claims. In truth, nineteen or twenty major earthquakes had ravaged the region in the past 10,000 years. Some of them took place at three-century intervals, others after more than 800 years. A probability estimate says nothing about how events are clustered.

When there is physical evidence that an earthquake is more likely during a certain period, things change. But Garry Rogers never said we had 'dodged the Big One' for now. A major quake might occur as I am writing these lines or while you are reading them, though with a smaller probability than during times of tectonic instability similar to the one his team discovered.

In spite of such misunderstandings, probabilistic forecasts are not as confusing as those contradicting each other. Sometimes we read headlines that say: 'Scientists claim earthquakes not predictable', and at other times: 'Experts warn of major quake'. Whom should we believe?

The Many Facets of Prediction

Contradictory warnings may not be as mutually exclusive as they seem. After all, the term 'prediction' can have various meanings. To soften the confusion, seismologists distinguish between 'prediction' and 'forecast', the first suggesting more precision than the second. So before trying to understand whether earthquakes are predictable, we must clarify what that means.

The public has high expectations. People would like to know *when* and *where* an earthquake will strike and *how intense* it will be. The timing should have an error of at most a few hours with a warning issued at least half a day in advance. The location has to be focused enough to avoid useless alerts. Finally, the intensity estimate must clarify whether an evacuation is needed.

Predictions that satisfy all these requirements have never been issued. Nevertheless, the experts can provide us today with some useful warnings. Certain regions, for instance, are more seismically active than others. A map outlining the history of tectonic shocks reveals areas that experience repeated earthquakes. With few exceptions, they correspond to the lines along which mountains, island arcs, and ocean ridges grow, such as the west coast of the Americas, the Aleutian Islands, and the mid-Atlantic ridge.

With the help of this map, seismologists issue forecasts using probability theory, but even their best results have large margins of error in time, intensity, or location. In 1985, for instance, some experts predicted a 95 per cent chance of a major earthquake before 1993 along the Parkfield segment of San Andreas's fault in California. Though small tremors occurred in the region within the predicted interval, none of them turned catastrophic. The experts had relied on the quasi-periodicity of earlier earthquakes, but their probabilistic approach didn't work.

This crude type of forecast can be taken further. Using more detailed information about tectonic movements in a specific region, seismologists make statistical predictions of the type Garry Rogers and his team issued for Vancouver Island. Thus they can warn of an increased probability for major earthquakes during certain intervals of time. The periods can range from a few days, as on Vancouver Island, to several years, as was done in China in the 1970s. Such warnings, however, bring no guarantees.

So far, no prediction has been as good as the one in Haicheng. Unfortunately the Chinese experts could not even come close to it the following year in Tangshan. Perhaps frustrated with this failure, they warned of a devastating earthquake in the Guandong region a month after the Tangshan disaster. Many people slept in tents for two months, but no major earthquake occurred. This error cast another doubt on the power of science to issue reliable predictions in terms of magnitude, time, and location.

What makes forecasting so difficult? Is there any chance to achieve better results in the future? To appreciate the answers to these questions, we will take a brief look at the first attempts to understand earthquakes and at how the accumulation of knowledge has led to the birth of seismology.

Beginnings

Earthquakes have been recorded since ancient times. Many sacred texts from all over the world mention them. The writings of early historians like Herodotus, Thucydides, Livy, and Pliny also describe major seismic events. But until the middle of the eighteenth century, very little was done to understand their cause.

An early but isolated contribution seems to have taken place in AD 132, when Zhang Heng, a Chinese philosopher of the Han Dynasty, invented the seismoscope. This instrument looked like a large wine jar with eight dragonheads stuck around its rim, each of them holding a ball in its mouth. When an earthquake struck, the direction of the wave made a dragon release its ball, thus providing some information about the shock.

Seismic research took a boost after the earthquake of 1755 that affected Portugal and Spain, triggering the Lisbon tsunami mentioned in the previous chapter. Two years later, John Bevis, a physician and amateur astronomer, published a collection of memoirs entitled *The History and Philosophy of Earthquakes,* in which several scientists presented their thoughts about the origins of seismic shocks. These works were based on recent records and not on ancient writings, as previous research had been.

Some essays were posthumous, like those of Robert Hooke and John Woodward. The rest belonged to no less famous living scientists. One of them was George Louis Leclerc, Compte de Buffon, known for his encyclopedic work on natural history. Buffon classified earthquakes by their origin into volcanic and others. Though he did not identify plate movement as the cause of 'others', his classification was correct.

The same year saw the publication of another work on earthquakes, *Memoirs Historiques et Physiques sur les Tremblemens de Terre*. Its author, Élie Bertrand, was a Swiss pastor and geologist, and a member of several European academies. The book focused on the origin of seismic movements and on the theoretical and practical tools for researching them. To justify his findings, Bertrand included data on several earthquakes, the one of 1755 among them.

Though making an important step forward from the descriptive work of their predecessors, these authors did not reach too far. But the time for more profound contributions had arrived, and seismology was on the brink of becoming recognized as a science. Its founder was the British geologist John Michell, who managed to explain how earthquakes occur and propagate.

Historians don't agree on whether Michell was born in 1724 or 1725, but they are unanimous about his genius. He studied at Cambridge, where he also taught after graduation. His lectures in mathematics, Greek, and Hebrew were so appreciated that the university offered him the geology chair named after John Woodward—his predecessor in seismic studies. Fourteen years later, however, Michell quit his professorship in favour of a rectory position in the country and got married. Nevertheless, he continued to pursue research in various directions, including earthquakes and astronomy. Happy to have found his peace of mind in a Yorkshire village, he led a quiet life until his death in 1793.

In 1760, while still in Cambridge, Michell published a treatise on earthquakes in which he laid the foundations of seismology. He didn't know that the Earth's crust was divided into plates, so he explained earthquakes through the force of underground steam. In his view, a small quantity of water vapour produced slight tremors, whereas large amounts led to seismic waves. So though he failed to find the intimate cause of earthquakes, he understood the principle of their propagation.

Michell's work was neglected for more than half a century although he had been well known in scientific circles. He counted among his friends top scientists like Henry Cavendish and William Herschel. But in spite of his connections, Michell received full recognition only in the second decade of the nineteenth century, when rave reviews of his memoir appeared in England. It took almost two more decades until a new generation of researchers caught up with his results.

These years of relative stagnation saw a significant addition of seismic data. Among the studied earthquakes were one in Calabria, Italy (1783), and two in Chile: Valparaiso (1822) and Concepción (1835). Notable in this period was also the contribution of the German mathematician and physicist P. N. C. Egen, who introduced the first magnitude scale. In his classification, grade 1 corresponded to barely perceptible tremors and grade 6 to fallen chimneys and cracked walls.

An innovative researcher of this period was the French mathematician and astronomer Alexis Perrey. Starting with the 1850s, Perrey worked in seismology for almost three decades. He was the first who tried to predict earthquakes, hoping to find some periodic factor that helped trigger them. The Moon's motion attracted most of his investigations.

Using statistics, Perrey announced three laws in which he related the probability of seismic events with the relative position of Earth and Moon. His second law, for instance, stated that earthquakes are more likely when the Moon is closest to Earth. The experts debated this gravitational approach for almost a century, but dropped it in the end for lack of evidence.

Contemporary with Perrey was the Irish engineer Robert Mallet, who worked for his father's construction firm. One day, while reading how the Calabrian earthquake had twisted a pair of pillars without overthrowing them, he noticed a serious flaw in the explanation of this phenomenon. His desire to learn more about the effect of shock waves on buildings made him get involved in seismology. The papers he published between 1845 and 1877 established him as one of the founding experts of the field.

A nineteenth-century historian of science wrote that Mallet had treated seismology 'in a more determinate manner and in more detail than any preceding writer'. Indeed, Mallet was in advance of his time. For example, he understood correctly that earthquakes are produced either by 'the sudden flexure and constraint of the elastic materials forming a portion of the Earth's crust, or by the sudden relief of this constraint'.

Mallet attached great importance to determining the velocity of earth waves. For this purpose, he conducted experiments in zones with various geological compositions, from the wet sands of Killiney Bay, at Dublin, to the granite country of the neighbouring Dalkey Island. By exploding charges of gunpowder, he measured that in sandy terrain waves travel twice as slow as through solid granite (825 versus 1,665 feet per second).

Mallet also compiled a comprehensive catalogue of recorded earthquakes, produced a seismic map of the world, introduced new methods of investigation, and did several detailed studies of contemporary earthquakes, such as the one that shook Naples in December 1857. Through his work, he also enriched seismology with numerous fundamental concepts.

After Mallet, the study of seismology began to spread. Researchers became active in several European countries, mostly in Italy, Germany, Austria, Switzerland, France, and Britain, as well as in the United States. They invented new instruments, the seismograph among them, and established the first seismological societies. With a foundation in place, this young branch of science could now develop.

In the more than a century that has passed since those early developments, progress has been huge, and seismologists have always directed their studies towards the final goal of making forecasts. But how far do they understand earthquakes today, and how does their knowledge guide them in their quest for predictions?

What We Know Today

Recently, while browsing a modern seismology textbook, I noticed the modesty of its claims. 'Though seismology provides a great deal of detail about the slip that occurs during an earthquake,' it read, 'we still have only general ideas about how earthquakes are related to tectonics, little understanding of the actual faulting process, no ability to predict earthquakes on time scales shorter than a hundred years, and only rudimentary methods to estimate earthquake hazards.' Then how could the Chinese experts warn about the Haicheng event six hours before it struck?

The full answer to this question will become apparent later. For now let us just remark that scientists are a conservative lot when it comes to their field, and most of them prefer to understate their results than to be blamed for overrating them. As Freeman Dyson, a British-born American mathematician and physicist, famously stated: 'The professional duty of a scientist confronted with a new and exciting theory is to try to prove it wrong. That is the way science works. That is the way science stays honest. Every new theory has to fight for its existence against intense and often bitter criticism.'

Like many other sciences, seismology uses mathematical models to examine how earthquakes occur and develop. The rupture triggered during an earthquake involves several physical actions, which lead to the propagation of various waves through the Earth's crust. Since most of these processes can be only guessed at, the models are simpler than the physical reality.

Seismologists see Earth as a viscous sphere with a thin peel that is broken into pieces. It would not be too far fetched to compare the crust with the skin that floats on the surface of a milk pot. Earthquakes are generated along the borders between patches when one patch pushes under or against the other. A first problem seismologists face is to understand how these patches move due to the flow of the viscous matter beneath. Then they focus on specific segments of each fault and try to describe the motion in detail.

The main problem with understanding tectonic movements is the lack of data. We don't know the precise location of most faults. Drilling the crust to find out is difficult and expensive. The deepest hole took twenty-four years to drill in spite of being only barely over twelve kilometres long. But even if we could go deeper, drilling alone would not suffice to give us a full picture of what happens under our feet; it would rather resemble the attempt to assess a person's broken bone with a thin and long needle.

Most knowledge seismologists obtain is indirect. Every new earthquake gives them the opportunity to gather data and improve their local models. Plate movements trigger waves, which travel through the Earth's mantle and crust, informing them about what's inside Earth. They have thus discovered that the crust is far from homogeneous, with depths that vary between ten and seventy kilometres. They know how the plates are distributed, have an idea of where they meet or overlap, and in what directions they move. Still, there are many unknowns about the physics of tectonic plates.

Progress towards understanding such phenomena is slow and costly. A recent project aimed at implementing sophisticated instruments 250 metres underground to measure the tremors noticed periodically on Vancouver Island came under the umbrella of a 300-million-dollar programme. Other research, which uses satellite monitoring, is several times more expensive. In spite of these high costs, the expected results are not even close to deterministic forecasts.

Consequently there is no consensus among experts about whether the time, location, and magnitude of major earthquakes could be determined before the event. Some think this goal should not even concern them. Prediction, they say, is mission impossible.

The Sceptics

Robert Geller is a professor at the University of Tokyo, a city where he has lived for more than twenty years. He specializes in computational seismology and is a strong believer in the impossibility of making deterministic earthquake predictions. In his opinion, we will never go beyond statistical forecasts, which can at best tell that a region is prone to major earthquakes every so many centuries. As a main reason for this limitation, he invokes the chaos phenomenon, which characterizes tectonic movements.

Not everybody agrees with Geller. Some seismologists claim that given enough data and computational power, they could make deterministic predictions, even if only in the short term. This limitation would create no practical problems. Warnings issued hours in advance would be good enough to evacuate the population from buildings.

Geller accepts the idea that chaos could be beaten with short-term predictions but only when you have a mathematical model that describes the studied phenomenon with good enough approximation and sufficient data to validate the theory. Unfortunately seismologists don't know the equations that describe how earthquakes occur, so they cannot develop a general model that would fit all cases. But Geller doesn't explain why these equations should be out of bounds forever.

Nevertheless, he admits that the discoveries of quantum mechanics are influencing his philosophy. Physicists have realized during the first half of the twentieth century that predictions are limited to the macroscopic level. At the quantum scale the notions of past, present, and future don't have the same meaning as in the world we perceive through our senses. Earthquakes, however, have nothing to do with quantum physics, so Geller's admission doesn't justify his belief.

'There is no particular reason to think earthquakes should ever be predictable in a deterministic sense,' Geller said in an interview given in 2001. This statement surprised me even more. It was like reading about a nineteenth-century physicist who saw no evidence why humans would ever land on the Moon. But Geller didn't stop here. On a different occasion, he went as far as to compare the attempts to predict earthquakes with those of turning lead into gold.

'No less of a scientist than Sir Isaac Newton regarded alchemy as his primary research field,' Geller wrote. 'His continual failures drove him to despair and led him to give up science for a sinecure as Master of the Mint. Sir Isaac Newton's failures notwithstanding, alchemy continued to attract the fruitless efforts of talented scientists for another hundred years. Earthquake prediction seems to be the alchemy of our times.'

Geller didn't explicitly ask for a halt to research in this area. Nevertheless, he thought that further work would give a better understanding of why deterministic earthquake prediction is impossible. In his opinion, there is no region of the Earth with a zero chance of tectonic movements. As an example he cited the New Madrid area of Missouri, where strong earthquakes took place in 1811 and 1812, although the region is apparently as safe as any other part of the Midwest.

Experts like Geller seem to draw their convictions from two sources: the empirical character of all the attempts to make predictions and the many erroneous warnings issued in the past. Between 1996 and 1999, for instance, China experienced thirty false alarms, which affected some local economies. No wonder that the Chinese government has put an end to the practice of unofficial forecasts.

Though Geller may appear as an extreme voice, his criticism could be viewed in the context of a widespread opinion among experts, which Susan Hough of the US Geological Survey in Pasadena, California, expressed in the following words: 'Seismologists would like to be able to predict earthquakes. If they criticize prediction methodology, their criticisms are driven not by an innate desire to see a proposed method fail, but by the need for rigor and care in this endeavor.' Indeed, without this balanced point of view, pseudoscientific practices might discredit their field.

Publicity Seekers

Foretellers of catastrophes have always existed. Some of them achieved lasting fame, such as Nostradamus—as the French Renaissance writer Michel de Nostradame is better known— whose prophecies are still reprinted almost half a millennium after he made them. Although false forecasts are easy to discredit post factum, apocalyptic claims continue to grip the attention of the masses. These scenarios seem both to fascinate and frighten those who choose to accept them.

In 1977 Charles Richter, the inventor of the magnitude scale that bears his name, expressed his annoyance with those developments. 'Since my first attachment to seismology,' he wrote, 'I have had a horror of predictions and predictors. Journalists and the general public rush to any suggestion of earthquake prediction like hogs toward a full trough... [Prediction] provides a happy hunting ground for amateurs, cranks, and outright publicity-seeking fakers.'

One such example is that of Iben Browning, a retired biologist who predicted in 1990 that a major earthquake would devastate the Missouri region of New Madrid. He gained credibility after *The New York Times* wrote he had correctly predicted the 1989 earthquake in San Francisco, and a major television network claimed his theory had been verified six times. The media frenzy escalated after 26 September 1990, when a modest 4.6 earthquake shook the region. Many schools closed and thousands of homeowners bought earthquake insurance. In the end, no catastrophe happened.

Browning's prediction was under the influence of James Berkland, an American geologist and editor of an online earthquake newsletter called *Syzygy*. The term, derived from Greek, denotes the alignment of three or more cosmic bodies. At eclipses, for instance, Sun, Moon, and Earth reach a perfect syzygy. Berkland, however, adopted a looser meaning of the term, calling syzygies even the positions that come close to alignments, like those Earth, Moon, and Sun have every two weeks.

According to Berkland, the composed gravitational pull of Sun and Moon at syzygy stimulates earthquakes. His theory is reminiscent of Alexis Perrey's eighteenth-century attempts, mentioned earlier, to link tectonic movements with the Moon's position. But Berkland is no more convincing than the French pioneer. Seismologists had noted that the Moon is too small to account for earthquakes. The Earth, which is eighty-one times the mass of the Moon, may cause quakes on its satellite, but not the other way around.

Berkland disagreed. Since earthquakes are chaotic, even small perturbations can trigger them. He gave two examples: the December 2004 Indonesian event, which led to the deadliest tsunami ever, and the March 1964 Alaska disaster, the second largest earthquake recorded in modern times, both of which had taken place at syzygies. These examples, however, prove nothing in the absence of some statistical clustering of earthquakes at the time of Earth–Moon–Sun alignments.

Similar gravitational theories had appeared before the 1982 'great alignment' of our solar system, a rare configuration that brought the planets within a right solid angle from the Sun. Ieronim Mihăilă, who was then my astronomy professor at the University of Bucharest, told me of a paper he had just published, which proved that the gravitational effect of this alignment would be negligible. His computations showed only a change of two millimetres in ocean tides. Indeed, earthquake activity was normal in 1982, and no geological disturbances occurred.

These historical facts, however, didn't impress Berkland. In addition to syzygies, he used other indicators, such as the number of missing dogs and cats, which—he had noticed—increases before major tectonic events. He claimed to have correctly predicted the October 1989 San Francisco earthquake based on this correlation. But Berkland's former colleagues of the US Geological Survey didn't support his theory. To them the matter is more complicated than that.

In fact, there are several Internet resources dedicated to earthquake prediction, none of which seems trustworthy. In an age when anybody can open a webpage, claims that try to impress the public appear like mushrooms after the rain. But the Internet only amplifies a trend that has always existed, and which often makes it difficult to distinguish between scientific and dilettante conclusions.

Unfortunately, top experts can also be wrong. An example is that of Bailey Willis, a distinguished professor at Stanford, president of the Seismological Society and the Geological Society of America, and a strong supporter of building codes. After the Santa Barbara earthquake of 1925, some cities adopted his safety proposals, but most of them ignored him. Frustrated, Willis warned of a major earthquake in southern California within ten years. The publicity that followed annoyed the powerful real estate and building industries, which lent a hand to discredit him professionally and humiliate him in the eyes of the public long before his prediction proved wrong.

Since then, western seismologists have been cautious. In the rare occasions when they go public about the likelihood of an earthquake, they attach a low probability to their forecast as the weight of evidence currently suggests. Nevertheless, they show no reservation when discussing whether deterministic predictions are possible. This issue is debated now more than ever.

The *Nature* Debates

In 1999 the journal *Nature* initiated a written debate entitled 'Is the reliable prediction of individual earthquakes a realistic scientific goal?' Between the end of February and the beginning of April, several researchers were asked to express their views on this question and to comment on the opinions of the others. If other experts wanted to join the discussion, they could do so by email.

The moderator was Ian Main, a professor in the Department of Geology and Geophysics at the University of Edinburgh in Scotland. With research interests ranging from earthquake mechanics and global tectonics to seismic and volcanic hazard, Main outlined the goals of the debate. He noted that the term *prediction* could be very confusing, so he divided the concept into four categories.

Main called the first type of prediction *time-independent hazard*. In this setting, the earthquake is viewed as a random process in time. Seismologists can use past earthquake locations and data about specific faults to compute long-term hazards. Such results, whose derivation poses no problem today, are useful for planning the best building designs and estimating the premiums of earthquake insurance.

The second type of prediction, called *time-dependent hazard*, accepts the idea that earthquakes are more likely at certain times than at others, as seems to be the case on Vancouver Island. As mentioned earlier, the probability of a major seismic shock is higher in this area during a certain week every fourteen months. Such estimates can be made in some parts of the world, and they allow better earthquake preparedness.

The third type of prediction, which Main called *forecasting*, uses precursors, and is still probabilistic in terms of magnitude, time, and location. A forecast would give the population months or weeks to get ready for a likely disaster. The difficulty, however, is to recognize the signals. An American-Indian research team attempted to identify a precursor in the area where I live: the algae blooming in the coastal waters in June 2006. In their view, the levels of chlorophyll, a pigment used in photosynthesis, go up a few weeks before an earthquake. But some of my colleagues at the University of Victoria were sceptical about this conclusion. They must have been right, for no major earthquake occurred.

Main called the fourth kind of prediction *deterministic* and asked that the magnitude, time, and location of occurrence be given with good approximation. While time-independent hazard is the standard prediction made today, the experts come closer to issuing time-dependent constraints. However, forecasting is rare, and some seismologists see deterministic predictions as an unreasonable goal, given present knowledge.

Ian Main ended his introduction with the question: 'So if we cannot predict individual earthquakes reliably and accurately with current knowledge, how far should we go in investigating the degree of predictability that might exist?' The first experts to respond were Robert Geller of the University of Tokyo, Max Wyss from the University of Alaska in Fairbanks, and Pascal Bernard of the Institut de Physique du Globe in Paris. While Geller expressed his characteristic scepticism, the other two made different points.

Wyss disagreed from the start with the idea that earthquakes strike suddenly and unexpectedly. Sometimes there are warnings we are able to recognize, as it happened in Haicheng in 1975, when the Chinese experts identified foreshocks and animal behaviour correctly. But if we don't see precursors or misunderstand them, it doesn't mean they don't exist. It may be that we are not ready to read them. 'I am pessimistic about the near future,' Wyss concluded, 'and optimistic about the long term.'

Pascal Bernard differed from both Wyss and Geller. He thought that seismologists should be primarily concerned with understanding the instabilities of the crust. The clarification of these basic aspects might eventually lead to better predictions. Good forecasts could be achieved when substantial progress is made in this direction. In other words: leave aside the prediction issue for now and try to grasp earthquakes better.

Andrew Michael of the US Geological Survey in Menlo Park, California, commented on the meanings of prediction and precursors. In his view, the efforts done to identify precursors had been immense, so perhaps other directions should be attempted. He expressed hope in the future and advised seismologists to heed the advice of Sir Peter Medawar, a 1960 Nobel Prize winner, according to whom 'no kind of prediction is more obviously mistaken or more dramatically falsified than that which declares that something which is possible in principle will never or can never happen'.

Several other experts entered the debate, some of them reserved, others with more open-minded approaches. Those who had started the discussion got the chance to respond to criticism. A frustrated Geller asked prediction supporters to 'refrain from using the argument that prediction has not yet been proven impossible as justification for prediction research'. Wyss noted that if prediction research were not funded, as Geller had suggested, no progress would be possible in this direction. The discussion seemed to slip towards the issue of which research area should get money.

A couple of email contributions brought some fresh ideas into the debate. Didier Sornette from the University of California in Los Angeles, for instance, pointed out that the seismological community has not brought its full potential to bear on understanding earthquakes and making predictions. Directions like statistical physics, super-computer modelling, and large-scale monitoring of physical measurements have been only marginally employed so far. In Sornette's view, it is premature to conclude that deterministic predictions are impossible. Like most of his colleagues, he expressed support for prediction research.

Ian Main concluded the debate by noting both a degree of consensus and one of controversy among the experts. He pointed out that although seismologists would have to continue their research on prediction, the public should not forget that 'it is not earthquakes themselves which kill people, it is the collapse of man-made structures which does most of the damage.' So in parallel to funding prediction research, governments should ensure the implementation of building and infrastructure codes.

Main's last remark reminded me of another earthquake, which I experienced on the morning of 28 February 2001. I was alone at home in Victoria, working in my study, which happened to be next to the main entrance of my house. When the tremors began at 10.55 a.m., it took me a few seconds to walk to the front lawn. Most neighbours were at work, and nobody else showed up outside.

My first thought went to my family, but there was nothing I could do for them. While staring helplessly on as our house was shaking, I noticed that my car was parked in the driveway, next to the chimney. The tremors lasted a minute or so. In the end nobody got hurt in Victoria, and my car survived the chimney threat.

That earthquake, which had its epicentre near Seattle, registered magnitude 6.8. On average, four such events strike our area every century. The tectonic shift took place about sixty kilometres under the surface, deep enough to spare the region of much damage. Nevertheless, I couldn't stop thinking of how crucial it is where you are when a major calamity hits.

Reasons to Hope

In the light of the *Nature* debates, we should not expect precise earthquake predictions in the near future. The 1975 Haicheng warning seems to have been a lucky coincidence. Most predictions made since have been false, and large earthquakes have occurred unexpectedly. Still, the Haicheng forecast was not singular. The Chinese experts were successful in four other cases in 1975 and 1976, but none of those shocks had large magnitudes. Though these successes are insignificant from the statistical point of view, do they point at a method for making meaningful predictions?

A paper published in March 2005 in the *Bulletin of the Seismological Society of America* sheds some light on the Haicheng events. In the late 1960s, the State Seismological Bureau in China had identified the Liaoning Province of Haicheng as a seismic-prone zone for the next few years. Consequently the region was carefully monitored. Several tectonic movements, including a land shift in the nearby Bohai Sea area, led in 1974 to a forecast for a major earthquake within one to two years.

An intense educational campaign kept the population on alert, so when some experts issued a warning for an imminent earthquake after registering several foreshocks, a combination of organized and spontaneous activity drove the people outside their houses. At 7.36 p.m., when the earthquake struck, most citizens were neither at work nor asleep. They stayed outdoors, though the prediction had given no precise time and magnitude. With good reason, the authors of the *Bulletin* paper qualified the success as 'a blend of confusion, empirical analysis, intuitive judgment, and good luck'.

The educational campaign played some role in the evacuation. In the months before the event, the authorities had distributed leaflets that outlined some basics about earthquakes. Thus the population noticed the foreshocks, one of which was significant. More than this knowledge, however, fear kept people outdoors.

The tragic Tangshan earthquake of 1976 did not strike out of the blue, as it is often stated. Precursors had been there too, but they occurred on a large area, so it was not possible to say with any degree of precision when and where an earthquake would strike or how strong it would be. Therefore nobody issued a prediction. The lucky mixture of Haicheng coincidences failed to happen in Tangshan.

In truth, it is difficult to tell whether a sequence of foreshocks will lead to a major earthquake. Only 6 per cent of small tremors are followed by a major event. Luckily, foreshocks are not the only known precursors. In Greece, for instance, a lot of work has been dedicated to the anomalous behaviour of the electrical field in the rocks. Unfortunately this research did not lead to firm conclusions. Other attempts involved the increase of hydrogen levels in soils and the infiltration of radon in wells, to mention only the indicators that seem to appear before many earthquakes. Most candidates for precursors are local, so they prove useless for general predictions.

If these precursors have a more fundamental cause, it is unknown. Also, should they have one, it's not clear whether it can be detected. Research done so far has been without definite results. Hopefully the experts will reach some conclusions once they enlarge their pool of data.

Though nobody can make precise earthquake predictions today, hope is not lost. Seismic movements are dynamical systems, which can be described (at least in principle) by some differential equations. It might be indeed very difficult to obtain these equations, but the nature of the problem cannot be compared to alchemy. At worst, earthquakes are part of a chaotic system, so long-term predictions will always fail. Short-term warnings, however, should be possible.

A prediction of only a few hours prior to the disaster would be good enough to allow evacuations. Should such predictions prove out of reach, even warnings issued minutes in advance may bring some relief. 'Real-time' warnings can be made even in the absence of precursors. Indeed, since light travels faster than seismic waves, once an earthquake takes place, signals can be sent and received tens of seconds before the shock hits adjacent regions. This issue is an active subject of research today because of the potential benefits it offers, which include stopping high-speed trains to prevent derailments and cutting the gas supplies that could ignite fires in broken pipes.

The further advance of science and technology might lead to approaches in prediction nobody has yet envisioned. Before the implementation of the Global Positioning System (GPS), for example, tectonic junctions were impossible to locate with precision. Now a large amount of the data seismologists collect is based on this recent achievement. Future unexpected inventions could bring the era of deterministic predictions closer to us than we hope.

Even if we are now limited to giving only time and location hazards, the study of earthquakes is useful for predicting other megadisasters. Since the time of pioneering research, geologists have linked certain seismic shocks with volcanic eruptions, noticing that tremors of the crust often occur days before the start of intense volcanic activity. Since the explosion of entire mountains, the rock bombardment that follows, the subsequent expulsion of ash, and the flow of lava can lead to significant loss of life, we will now focus on understanding how such events can be forecast.

3. CHIMNEYS OF HELL: VOLCANIC ERUPTIONS

The cloud was rising from a mountain... into the sky on a very long 'trunk' from which spread some 'branches'. Some of the cloud was white; in other parts there were dark patches of dirt and ash.

Pliny the Younger

In the summer of 2000, I organized a session in celestial mechanics at the Third World Congress of Nonlinear Analysis held in Catania, Sicily. On my way from Rome, I stopped in Pompeii for a short visit. The place impressed me quite a bit. I had never thought that an ancient city could have looked so much like today: paved streets with sidewalks and bridges; stores with large signs whose letters were still visible; homes with tile floors, murals, bathrooms, and gardens—many of the houses well preserved.

Watching Vesuvius, I tried to imagine what had happened there in AD 79 when the volcano ejected enough ash to cover the city and kill those who couldn't flee. The events Pliny the Younger had recounted in a letter to Tacitus came to life in my mind. I could 'see' that immense cloud growing—its heat burning the breath of the people—and the rain of dust and rocks falling from the sky. The description of the Apocalypse must have been inspired from a similar experience.

In Sicily I stayed in Acireale, a seaport near Catania, at a beach hotel built on volcanic rock. Etna dominated the landscape. This view aroused in me the desire to go up the mountain, whose peak the Emperor Hadrian had attempted in AD 125 without success. I climbed there after the conference. Signs of a recent eruption loomed. The lava had flown down the slopes a few years earlier, destroying farms, houses, and roads in its path, finally freezing into forms that looked like surrealist sculptures.

Back home, I began to view Mount Baker—an ice-covered cone visible from Victoria—with different eyes. My trip to Italy had made me grasp that I also lived in a region with active volcanoes. Mount Saint Helens, which lies 300 kilometres south of my city, had erupted in May 1980. The stories I heard about that event piqued my interest in the volcanic history of this region.

If before studying the topic, I had thought that volcanic activity rarely made many victims, I soon changed my mind. 'During the past 400 years,' write Zeilinga de Boer and Donald Sanders in their book *Volcanoes in Human History*, 'perhaps a quarter of a million people have been killed as a direct result of volcanic eruptions. Indirect aftereffects, such as famine and disease, may well have tripled that number.' Some of the most devastating events of this kind took place only a few generations ago.

Mountains that Explode

The Malay Archipelago is a large group of islands dividing the Indian and Pacific Ocean between Australia and Southeast Asia. Situated along the border of two tectonic plates, the region has a

strong seismic activity. Indeed, an earthquake in this area triggered the December 2004 tsunami, and the two most famous
nineteenth-century volcanic eruptions happened here as well.
The first took place in 1815 on the Indonesian island of
Sumbawa, where the explosion of Mount Tambora was heard
2,000 kilometres away. According to the newspapers of the
time, some 50,000 people died in the event. The volcano ejected
sixteen cubic kilometres of ash and dust, which reduced the
transparency of the atmosphere and lowered the global average
temperature by 1 degree Celsius in the following thirteen
months. No wonder that 1816 became known as 'the year
without a summer'.

The first thing to be noticed far away from the eruption were
some luminous effects: the Sun appeared green in some parts of
Asia; evening skies in England turned red, yellow, pink, and
orange; and on certain days in New York the sunspots became
visible to the naked eye. Other consequences were less colourful,
such as crop failure, which led to high food prices. This development shook the economy and created a subsistence crisis that
triggered immigration waves within North America.

The dimming of the atmosphere and the economic downturn
seem to have also had a psychological impact. Many people
complained of depression, perhaps because of lacking enough
sunshine, and some gloomy literary creations, such as Mary
Shelley's *Frankenstein* and Lord Byron's poem *Darkness*, were
written. One way or another, the eruption of Tambora affected
every living human being.

The second eruption was not as strong as the first, but its
consequences were felt more intensely and could be better
evaluated. It happened in August 1883 some 1,100 kilometres

west of Tambora on the island group of Krakatoa, whose geography was altered after the event. The series of massive explosions was heard as far as the island of Rodrigues, near Mauritius, some 4,800 kilometres away, and produced a tsunami that reached heights of up to forty metres.

More than 36,000 people died, with 165 towns and villages completely destroyed and 132 seriously damaged. Part of Krakatoa vanished in the explosion, but it grew back during several quiet eruptions in the past century.

A consequence of the 1883 event was that of three tidal surges registered hours apart after the explosion in remote spots such as the English Channel and San Francisco. The original explanation of this phenomenon was that the tsunami propagated several times around the world along the shortest marine routes. But numbers didn't match. The water elevations occurred long before the wave could have reached those places.

The correct justification of this effect was given only in 1969 in the maiden paper of a young British physicist, Chris Garrett, then at the University of California in La Jolla, now a colleague of mine in Victoria. His computations showed that the water surge had been generated by the airwaves created after the explosions. The shocks had been so strong that the atmospheric pulse circled the globe several times, affecting the surface of the oceans during their first three passages.

Krakatoa had a stronger impact on scientific research than Tambora. The experts were better prepared to understand the consequences of major eruptions than two generations earlier. Field volcanology had become established, and some researchers made the trip to Indonesia to extract the secrets of the phenomenon. But the birth of volcanology as a modern science came only after another disaster rocked the world. This time it happened on the island of Martinique—a French colony in the Caribbean.

The 8 May 1902 eruption of Mount Pelée destroyed the city of Saint Pierre, known in Europe as 'the Paris of the West Indies', killing almost 25,000 people. Only two men survived in the city and a handful in the outskirts. A plug of molten rock had sealed the vent of the volcano, and the pressures ejected a glowing cloud of dust and gases at aviation speed, mostly in horizontal directions. Saint Pierre was six kilometres away.

This catastrophe didn't come out of the blue. There had been plenty of warnings during the preceding two weeks. Repeated eruptions of volcanic ash and stone, loud detonations, earthquakes, pillars of black smoke, the swelling of the rivers and their uprooting of trees and rocks, all had given good reasons for taking measures to prevent a disaster. But instead of ordering a safe evacuation, the regime discouraged it. The governor viewed the incoming elections and the existing political tensions as more important than the safety of the people.

An amazing story of survival is that of a young girl named Havivra Da Ifrile. Before the main eruption she looked down the vent of a crater located halfway to the top and saw 'the bottom of the pit all red, like boiling, with little blue flames coming from it'. She ran to the shore, jumped into a boat, and rowed into a cave just when the eruption started. The last thing she remembered was the water rising rapidly in the cave. A French cruise ship found her unconscious, drifting out to sea in her broken boat.

Mount Pelée remained active for three more years. It captured the attention of the media, and public awareness about eruptions reached unprecedented levels. The volcano had another period of activity from 1929 to 1932, but it has been quiet since. The lessons learned in 1902 had been helpful in those subsequent occasions.

The events in Martinique also influenced the decision taken about the location of the Panama Canal. In 1902, the US Senate was debating where to enable a new route between the Atlantic and the Pacific across Central America, with Nicaragua as a favourite candidate. But those who lobbied for Panama pointed out that Nicaragua was a country with active volcanoes, whose one-centavo postage stamps depicted a smoking mountain cone. They presented the entire Senate with those stamps, and won the decision by eight votes.

Most important for the development of science, the catastrophe at Mount Pelée gave a boost to volcanology by attracting many new minds to the field. These young researchers wanted to understand how eruptions happen. Some even aimed to predict them. But to appreciate the importance of this moment in the history of volcanology, we must take a look at how this new research field came into being.

Inside the Crater

Though volcanology had become recognized as a modern science only at the beginning of the twentieth century, humans have always been fascinated with eruptions. Legends from various cultures, such as the ancient Greeks, the Athapascans of British Columbia, and the Polynesians of Tonga Island, attribute fire to the depth of volcanoes, whose name comes from Vulcan, the Roman god of fire. Historians like Thucydides and poets like Ovid mention volcanic eruptions too.

The early philosophers already had some theories about volcanism. While Pythagoras believed in the fire hypothesis, Empedocles thought the Earth's centre to be composed of molten matter, whose rise led to explosions and flows of lava. Aristotle tried a physical explanation, stating that the heat generated through the friction of the underground wind and the solid matter created the eruption. Seneca proposed sulphur and other flammable substances as potential heat sources.

The Middle Ages marked a drawback as the Catholic Church banned most ancient science. Volcanoes became 'chimneys of Hell', and any attempt to understand them was reduced to religious doctrine. The idea of combustion as the cause of eruptions returned with the Renaissance, but it was often mixed with alchemy and mystical explanations, which vanished gradually in the sixteenth and seventeenth century.

Signs of a turn occurred when some thinkers, René Descartes among them, began to regard the Earth as a cooling star and considered its primordial heat as the cause of volcanic eruptions. Georges-Louis Leclerc, Compte de Buffon, subscribed to this theory and took it further. He computed the Earth's age, starting from the current surface temperature and taking into consideration the cooling rates of metallic spheres. He thus wrongly predicted that our planet would completely freeze within 100,000 years.

In the second part of the eighteenth century, geologists initiated fieldwork to study volcanoes. One of the pioneers was William Hamilton, a British aristocrat born in 1730. As a Member of Parliament, he became the British Envoy to the court of Naples, where he spent thirty-six years. In his spare time he studied Vesuvius and published many observations about its volcanic activity.

Hamilton determined the chemistry and mineralogy of volcanic rocks, hoping to understand how volcanoes erupted. He differentiated between *lava* and *tufa* (or *tuff*) rock, the latter originating from volcanic ash, and being commonly used in Italy as a building stone. Living close to Pompeii, he also gathered a large archaeological collection, which the British Museum bought to develop its Greek and Roman section.

Hamilton also distinguished between quiet and noisy eruptions. He noticed that explosions occur only when water infiltrates inside the cone and comes in contact with the hot magma, which is the viscous substance formed within the crater from melted rocks. Otherwise, lava—the expelled magma—spills quietly, as often happens in Hawaii.

Hamilton was not alone in his fieldwork. His French contemporary Jean-Etienne Guettard also made important contributions. Guettard studied botany and medicine in Paris and became the doctor of the Duke of Orleans, whose large rock collection stimulated his interest in geology. With his financial future secured by a pension he received after the duke's death, Guettard dedicated most of his time to volcanoes.

In 1751 he took a trip to central France. A friend joined him. They stopped on the way in the Auvergne region, where they saw some unusual stones in house walls and road pavements. The rock was black and porous and looked very much like what they had seen years earlier while hiking the slopes of Vesuvius together. This observation led them to the quarry and the discovery of seventeen craters which had been active more than 7,000 years earlier. The study of this region would be fundamental in the birth of volcanology.

Another French geologist who deciphered the rock structure of the Auvergne region was Nicholas Desmarest. He was the first to understand that basalt had volcanic, not marine, origin— as Guettard had concluded. This observation, made by tracing some columns of dark basalt to their source, had an important impact on unravelling the mysteries hidden under the Earth's crust. It also emphasized the role of volcanic eruptions in shaping valleys and riverbeds.

Italians didn't lack interest in volcanoes either. Perhaps their most important researcher of this period was Lazzaro Spallanzani, a professor at the University of Pavia, who discovered the key role of sulphur in the melting and bubbling of volcanic rocks. He also showed that sulphur plays no role in the formation of certain glassy rocks, thus refuting earlier beliefs.

Spallanzani rejected Aristotle's explanation that eruptions are produced when the wind blows underground through cracks. But in spite of understanding eruptions better than his ancient predecessor, he had his limitations too. He considered, for instance, heat as a fluid substance, which is pushed up with the magma to the surface, where it vanishes when the lava cools down. The idea that heat is a form of energy resulting from fast molecular motion had still to be born.

These early contributions had come a long way from the belief in the 'chimneys of Hell' that had dominated the Middle Ages. The earth sciences, however, had not yet evolved enough. It would take a few more decades until they would branch out and grow into specialized fields. And though forecasts of volcanic eruptions were still far away, a German researcher tried to get there through his imagination.

Dreaming of Predictions

In 1755, at the age of 18, Rudolph Erich Raspe registered at the University of Göttingen. A few weeks after classes began, he learned about the huge earthquake and tsunami that had taken thousands of lives in Portugal on 1 November. The news strengthened his determination to study earth science and understand these phenomena. His interest in rocks had been with him since childhood due to his father, who worked for the Hannover Department of Mines.

After graduation, Raspe began an intense research activity. His published work attracted attention right from the start. One of his studies was about the birth and growth of volcanic islands. He argued that they result from the uplift of the ocean floor, and that lava only covers their surface. These ideas stemmed from some forgotten work of the English scientist Robert Hooke that Raspe had revisited.

The young German scientist began to correspond with famous colleagues, William Hamilton and Benjamin Franklin among them. Raspe met the latter in 1766, when the man who would soon become one of the founding fathers of the United States visited Germany. These developments enhanced Raspe's reputation, helping him secure the position of secretary at the Royal Library in Hannover, where he pursued theoretical research and took fieldtrips to study the volcanic origin of basalt.

But Raspe's private life deteriorated because of some bad investments. To escape ruin, he resorted to embezzlement and later married a wealthy woman. When his dishonesty was discovered years later, he abandoned his family and fled the country. Raspe ended up in England, where he was well received in the circles of the Royal Society, but the doors closed to him when the news of his criminal acts reached London.

Though penniless and dishonoured, Raspe didn't give up. He soon published a successful book on the progress of German geology. This work, however, did not secure him an academic position as he had hoped, so he was forced to make a living as a German translator. In the end he found a job as a mine prospector in Cornwall, where he began to write fiction in his spare time.

In 1781 an anonymous author published a German collection of stories about Baron Münchausen. Raspe rewrote and expanded them in a volume that appeared in English in 1786. Other authors continued to publish sequels throughout the nineteenth century, making this comic character hugely popular in Europe and, to some extent, all over the world. Indeed, I remember enjoying a weekly TV show featuring him during the 1960s, while I was growing up in Romania.

Raspe's character shows interest in volcanoes. In chapter 20 of *The Surprising Adventures of Baron Münchausen*, he climbs Mount Etna and descends into the crater, where he finds himself in the company of Vulcan, the Roman god. There he learns that the volcano is nothing but 'an accumulation of ashes thrown from his forge', and that 'the appearances of coal and cinder in the world [which] mortals call eruptions' are nothing but his quarrels with the Cyclopes. Vulcan also assures him that 'Mount

Vesuvius was another of his shops, ... where similar quarrels produced similar eruptions.' By making the baron guess the mood of his host, Raspe implies that eruptions are predictable.

The German scientist never took his fiction seriously. He kept prospecting British mines, unaware that Münchausen would secure him a place in the history of literature. In the fall of 1794, while working in Ireland, he fell ill with a fever and died shortly thereafter. He was buried in an unmarked grave in a chapel graveyard at Killarney.

Hazard Zones

During the nineteenth century, geologists understood volcanoes better and could describe the mechanism that leads to the spilling of lava. Thus they learned that two main factors trigger eruptions: high underground temperature and pressure. Heat melts the rocks and creates magma, which rises due to pressure and makes its way up to the surface through fissures and vents.

The fluidity of magma is responsible for the difference between explosive and silent eruptions. When the mixture is more fluid, gases escape easily, pressure builds up, and the magma blasts into the air, carrying ash and rocks with it. Magma with a higher degree of viscosity comes out of the earth slowly and flows down the mountain.

Lava rarely kills people because they have time to run out of its way, but leads to material damage, and occasionally buries entire communities. Still, sometimes lava can be fast, with speeds exceeding sixty kilometres an hour. More perilous, however, is the ash, which can cover cities—Pompeii being the best-known example. The most dangerous, of course, are the explosive eruptions. Thousands of people might die under such circumstances, as happened at Tambora, Krakatoa, and Mount Pelée.

The last two events occurred at a time when science was ready for a breakthrough. The publicity surrounding them stimulated the work that led to the birth of modern volcanology. Thus, at the beginning of the twentieth century, predictions didn't seem unrealistic anymore. And unlike earthquakes, which are so difficult to forecast, the volcanoes revealed many of their secrets.

But forecasts came at a price. The experts had to monitor volcanoes and track their history of eruption. So the first thing they did was to identify the hazard zones that had been active in the past 10,000 years. Today, the number of dangerous volcanoes is about 1,500. Some are in the Mediterranean, others in South Asia, and approximately 75 per cent of them form the Pacific Ring of Fire, which stretches from the west coast of North and South America to New Zealand and East Asia.

The Ring of Fire is not continuous. Only certain segments are dangerous. Some of the volcanoes have familiar names: Mount Saint Helens in the United States, Popocatepetl in Mexico, Cotopaxi in Ecuador, Krakatoa in Indonesia, and Mount Fuji in Japan. But like Paricutin, a young volcano, others will appear in the future along the junctions of Earth's tectonic plates, and yet others will grow extinct. Volcanoes stay active between one and two million years, which is quite a short lifespan on the geologic time scale.

Once the hazard zones are mapped, the experts consider each region separately to determine its characteristics. Though all volcanoes exhibit some general features, the models that allow prediction must be developed in each case, taking into account the local attributes. Thus volcanoes are classified in several categories, such as Hawaiian, Strombolian, Vulcanian, and Plinian, depending on their particularities.

Small earthquakes usually precede an eruption. People might not feel them, but seismographs do. The rising of the magma is signalled by a shift from deeper to shallower tremors. Still, seismic movements are not enough to predict an eruption. They may only signal its likelihood. The mechanism that leads to an explosion is complicated, and the experts need to develop mathematical models to understand it. So who are the scientists working in this direction, and what exactly do they do?

Seeking Scientific Formulas

Stephen Sparks is tanned most of the time in spite of living in England. As a volcanologist at the University of Bristol, he owes this complexion to his many fieldtrips all over the world where volcanoes are active. In July 1995 he learned about the eruption of Soufrière Hills on the island of Montserrat in the Caribbean. This event stirred wide interest, and Sparks was among the scientists who went to research it. During the eruption, they gathered large amounts of data that made them change their understanding of this volcanic mountain.

They had originally believed in the existence of a magma chamber, which cooled and crystallized beneath the volcano during its dormant periods. When an eruption began, the magma either exploded or flowed as lava. The results of their research, however, showed that this model was wrong. No chamber lay under the volcano, and the lava was a mixture of old and fresh magma, the latter coming from much deeper levels than previously thought.

The first confirmation of this new hypothesis came from lumps of dark basaltic rock in the lava. Such blobs signal depths where temperatures are higher than near the surface. Sparks had additional reasons to accept this view when his colleagues Jon Blundy from Bristol and Kathy Cashman from the University of Oregon found similar components at Mount Saint Helens, for which an analogue theory emerged.

Observations didn't suffice. Sparks needed a mathematical model that would explain the dynamics of the eruption under the assumption of a deep source. In particular, the model had to show that the magma could beat gravitation and rise high enough, prove the formation of gas bubbles, the occurrence of crystals, and all the other changes in the composition of the magma. This was no easy goal, but he had to try.

Sparks teamed up with Oleg Melnik from the Institute of Theoretical Mechanics at Moscow University, a mathematician with interest in volcanic flows. Together they developed a model based on the differential equations that describe the motion of a viscous, gas-saturated fluid in a highly pressurized environment. More precisely, they considered a tube some thirty metres wide and five kilometres long in which there is a fluid of crystals and molten silicate rock mixed with dissolved gas.

To solve these equations, they assigned the mixture an initial temperature of 850 degrees Celsius and imposed certain conditions, such as a drop in pressure at the tube's high end. The numerical simulations showed that the dissolved gas escaped and the fluid turned into solid matter, whose melting temperature rose to 1,100 degrees Celsius. Thus the viscosity of the mixture increased a billion times, blocking the tube a few hundred metres below the top.

Depending on the pressure that builds up under the blockage and the nature of the material involved, several things could happen: earthquakes could occur, the mountain could explode under pressure, or the blockage could be pushed up gradually, releasing the pressure and letting the lava flow smoothly out of the vent. More complicated scenarios, such as eruptive cycles, which are common for many volcanoes, could also take place. The model of Sparks and Melnik explains these phenomena and allows predictions if the initial conditions are known.

To measure the conditions inside the vent, the experts had invented in the 1970s an electronic device called a *tiltmeter*, designed to record tiny changes in angle when the volcano swells due to pressure increase. At Soufrière Hills, this instrument registered cyclic changes, in accord with the theory. These results allowed the prediction of the relief periods, useful for managing the crisis on the island of Montserrat. Telephone technicians, for instance, could restore the lines during the hours of respite.

Of course, models such as this one cannot predict everything, and they work only for volcanoes that behave like the one at Soufrière Hills. But once the basic equations are understood, progress follows. Luckily, volcanic prediction, unlike earthquake forecasting, has reached the stage of writing down these equations.

Other researchers took a different approach at Soufrière Hills. In September and October 1997, while volcanic activity was still intense, a team led by Charles Connor from the University of

South Florida in Tampa used a probability model to determine the repose intervals between two eruption sequences. The team assumed that the likelihood of renewed activity increases exponentially with the time passed since the previous eruption. They also included additional factors, such as gas loss, which helped them refine their estimates.

When enough repose time passes and no new sequence begins, it means that the volcano has ended the current eruptive cycle. In Montserrat, for example, the repose intervals lasted between two and thirty-four hours. The team computed that if forty hours pass without activity, then the cycle has very likely ended. However, to compensate for any margin of error and stay out of danger when climbing the mountain for fieldwork, they used a more conservative equation, which indicated that they should wait for eighty-five hours.

Volcanologists employ many mathematical models for various levels of prediction, from estimating the start of an eruption and the repose intervals to deciding when a cycle has ended. But models alone are not enough. Monitoring volcanoes and determining their characteristics is also important for prediction. Instruments can sometimes make real-time warnings, as it is done to avoid tsunami disasters. A useful procedure of this kind has been developed to trigger the alarm for other dangers that loom on volcanic slopes.

Lahar Detection

The residents of Seattle are proud of their city. Nature has been generous with them. The calm seawaters, the islands with evergreen trees, and the mild climate make this place ideal for nature lovers. From walks in the park under the corona of Douglas firs to kayaking and helicopter skiing, outdoor activities define this part of the world around Mount Rainer, which dominates the landscape with its frozen peak. But few know that this is not a safe zone. Among the natural dangers looming here are the lahars.

A word of Javanese origin, *lahar* describes a mudflow resembling a river of fluid concrete coming down the slope of a mountain. Lahars are mixtures of ash and water that develop from volcanic eruptions, heavy rainstorms, steam explosions, or snow melting. They can move huge quantities of debris in a short time, destroying everything in their path. The deposits they form at rest can sometimes be hundreds of metres thick.

More than 100,000 homes in the Puget Sound area of Seattle are built on lahar sediments. The largest deposit, on which several towns have grown, formed 5,700 years ago. The most recent is six centuries old and between 5.5 and 27 metres thick. Unfortunately some inhabited areas around Seattle could experience lahars anytime.

Nevado del Ruiz is a 5,389-metre-high Colombian volcano with an ice cap and a long history of lahars much like that of Mount Rainer. On 13 November 1985, a glacial outburst from a small eruption triggered a mudflow that came down the Azufrado valley. In four hours, the lahar covered more than 100 kilometres. Hardest hit was the town of Armero, with three quarters of its inhabitants killed. The total casualties mounted to more than 23,000 deaths and 5,000 injured, with 5,000 homes destroyed along the Chinchiná, Gualí, and Lagunillas rivers.

Survivors' accounts mention several waves of flowing material. The first, which arrived at 11.25 p.m., was an insignificant amount of dirty water the lahar had dislocated from a lake located upstream. The second came ten minutes later and proved to be the most destructive. Flow depths ranged between two and five metres. The third wave struck at 11.50 p.m. with a speed about half the previous one. The next six to eight pulses had no significant impact, with the last one registered around 1 a.m. The interviewed survivors caught in the mud recount how terrifying each new wave was for them.

Unfortunately Mount Rainer and Nevado de Ruiz are not the only volcanoes where lahars can occur any time. Other parts of the world are prone to such disasters, with Mount Ruapehu in New Zealand and Galunggung in Indonesia among them. The former was recently struck by a lahar, but luckily without victims. This event showed how much technology had progressed since the Armero disaster.

Indeed, the sad Colombian experience gave scientists food for thought. Keeping observers in place in such areas would be expensive and inefficient because highly destructive lahars happen rarely. But since many such slopes have been populated lately, and it is not clear which lahar will develop into a major one, the necessity of implementing a warning system is an important issue for the residents living in endangered zones. Luckily, experts from the US Geological Survey have recently developed an inexpensive and efficient programme for those areas.

The system is based on several stations placed down the volcano, each of them endowed with a seismometer that registers the vibrations produced by a lahar, a processor that analyses the signal, and a device that sends and receives radio signals to and from the base. A solar panel and batteries power the network.

Every half-hour, each station sends information about the vibrations to the main computer. If their amplitude exceeds a certain threshold, an emergency message is sent repeatedly for as long as the amplitude remains high. The frequency of the vibrations allows experts to distinguish between lahar-generated tremors and those triggered by other factors. For example, volcanic eruptions create tremors of less than six oscillations per second, while lahar vibrations are between five and twelve times more frequent.

This detection system is now implemented on volcanoes in Ecuador, Indonesia, Japan, Mexico, the Philippines, and the United States. A similar network installed on Mount Ruapehu by the New Zealand Department of Conservation succeeded in averting a catastrophe when a lahar came down the volcano on 18 March 2007. Images of the event were broadcast worldwide on television channels. Though nothing was said about the procedure, this warning was a success for all those who try to predict disasters.

Predicting Eruptions

An ideal prediction tells exactly when the volcano will erupt and allows time for determining the risk zone and organizing an evacuation. Such precision might never be reached, but vol-canologists try to get as close to it as possible. They divide their objectives in two categories: long-term forecasting and short-term prediction.

Long-term forecasts don't aim to be precise. They only give an idea about how often a volcano erupts and what to expect from an eruption. For this purpose, the experts study the geologic history of the volcano and assess the phenomena that have occurred in the past. Specifically, they examine several layers of rock deposits on and at the bottom of the mountain, use radioactive age dating, and determine the likely dangers: explosions, lava or lahar flows, gas poisoning, ash falls, and so on.

This information is shared with public officials, who consult the volcanologists to plan evacuations, rescue, and recovery operations. The authorities don't want to order evacuations that are larger than necessary. Their reluctance results not only from the risk and cost such an enterprise carries but also because not everybody is willing to obey. Prior to the 1980 eruption of Mount Saint Helens, for instance, an 83-year-old resident named Harry Truman, who lived in a cabin at the base of the volcano together with his sixteen cats, became the focus of the national media for refusing to leave his home. He died there the day the mountain exploded.

If long-term forecasts have their stage-setting role, short-term predictions inform the authorities when to act to protect the endangered population. Volcanologists have therefore focused on developing methods of predicting eruptions a few days in advance. For this purpose, they determine precursor events, which indicate when the magma will rise to the surface.

A first method used for this purpose is that of seismic exploration and monitoring. Seismic waves don't travel through fluids, so to determine whether magma exists in a certain zone, the experts produce seismic waves with small explosions. They identify the magma pockets by mapping the areas where the waves don't pass through. The earthquakes that take place before an eruption help them trace the rise of the magma.

Another method uses the resistance of rocks to the electrical flow, which depends on temperature and water content. Therefore electrodes placed in the ground to measure the change of a physical quantity called *specific electrical resistance* or *resistivity* may allow tracking of the magma.

Rocks also contain magnetic minerals. Above a certain temperature characteristic to each mineral, however, the magnetic properties vanish. (The threshold of magnetite, for instance, is 500 degrees Celsius.) The monitoring of the magnetic field may therefore indicate the whereabouts of the magma. An even more refined method measures the changes in the heat flow through remote infrared sensing.

The ground water system becomes unstable as the magma rises in the volcano. Consequently, water levels and temperature change, so measuring these parameters in wells also helps with the prediction.

Instruments like the tiltmeter used at Soufrière Hills provide information about the ground deformation that occurs when the magma fills the vent of the volcano. Other devices measuring the distance between several points on the mountain can also indicate the swell. These changes, however, are small. Consequently the instruments used for their detection must be highly sensitive.

The composition of the gases emitted from volcanic vents is another predictor. Prior to an eruption, the proportion of sulphur dioxide and hydrogen chloride relative to water vapour increases in general, so the detection of this change may help.

In practice, all these methods are attempted because not all the phenomena described above take place simultaneously before every eruption. The precursors vary with each volcano.

Mount Saint Helens provides an example of how monitoring precursors leads to prediction. After the 1980 explosion, a volcanic dome began to grow in the crater following small sporadic eruptions. The research of this phenomenon showed that the most sensitive indicators of approaching activity were the slight expansion and tilt of the dome and an increased release of sulphur dioxide. On 12 March 1982, the experts warned of an eruption within the next ten days. On 15 March they narrowed the window to four days, and on 18 March to forty-eight hours. The eruption took place the next day.

Not all predictions are as successful as the one just described. The variety of volcanoes and the specifics of each mountain often make forecasting difficult. The chaos phenomenon governs eruptions too. A bit of excess in sulphur gases, for instance, could transform an apparently harmless eruption into a catastrophe. Volcanologists are therefore careful when assessing the risk. For this reason, the US Geological Survey has put in place a colour code to indicate hazard levels, which provides a less precise warning than a specific prediction. Thus code green labels the normal non-eruptive state of a volcano, code yellow the elevated unrest above the known background activity, code orange the escalating activity with increased potential of an eruption, and code red warns that an eruption is imminent or underway.

The volcanoes known to cause trouble are closely watched. But sometimes eruptions happen where nobody expects them. A famous case occurred in central Mexico during World War II.

Out of the Cornfield

On a clear afternoon in February 1943, a Tarascan Indian farmer named Dominic Pulido joined his wife and son, who were watching the sheep pasturing on their land.

'Anything new?' he asked. In the past two weeks they had felt strong earth tremors in the region, and didn't know what was happening.

'Yes,' answered his wife Paula. 'We heard noise and thunder underground.'

It made no sense. The day was beautiful and still, with no cloud in sight. But a few moments after Paula had spoken, Dominic heard thunder too, as if they were in the middle of a storm.

Around 4 p.m., Dominic was making a fire when he noticed that a grotto situated on his cornfield had opened. He went to check it and saw a fissure about half a metre deep. As he returned to the fire, the earth vibrated and thundered, and the hole swelled, while a grey dust burst through the crack. He heard a loud whistling noise and smelled rotten eggs in the air. This occurrence terrified the Pulido family. It must be the Devil's hand, they thought.

By the next day, a cone—the size of a tall house—had grown, and it continued to send ashes and rocks up in the air. Though this was the only crack in the ground, the field was nothing but a mass of molten rock. The villagers believed they were witnessing the end of the world.

The Pulidos were the first people who saw the birth of Paricutin—a new volcano some 300 kilometres east of Mexico City. By 1952, when the activity ceased, its cone measured 424 metres. By this time, many farmers had lost their homes and land. Some left Mexico to seek work in the United States. And more than 1,000 people died following a 1949 eruption, apart from those killed because of related consequences.

Indeed, many villagers fell prey to diseases from the pollution following eruptions or due to the toxic gases. As their predator birds changed their migration patterns, some bugs that thrived in the new conditions spread diseases too. Still, it is difficult to assess all the collateral effects because they often mix with other factors. But there were some positive aspects too, like the enhancement of the soil, which yielded good crops and pastures. The quality of the soil is a well-known asset around active volcanoes.

Although events like the one that gave birth to Paricutin are rare and much harder to predict than regular eruptions because of lack of monitoring, raising awareness about this possibility would help prevent disasters. An educated population could alert the authorities about thunders in the ground and enable an evacuation if the experts decide it necessary. As the example of Tilly Smith—the girl who saved many lives on a Thai beach—suggests, disaster preparedness should become part of every educational system.

In the Line of Duty

When I climbed Mount Etna, I noticed signs that warned of not going beyond a certain point, some three kilometres away from the top. Since my curiosity was stronger than my caution, I ignored them. Later I understood that I had made a mistake. Etna was in a state of code yellow alert.

Volcanologists are aware that their profession can be dangerous sometimes. In general, however, they take safety measures and don't climb volcanoes when eruptions are imminent. Still, sometimes accidents happen. Some volcanologists died in explosions, others were killed by rocks or gas, and there were those who got caught in lava flows. At high risk are the expeditions that try to record eruptions—a noble activity, which helps not only with predictions but also with educating the public and raising awareness about the dangers volcanoes pose.

Two French pioneers in filming and photographing volcanic eruptions were Maurice and Katja Krafft. They met in the 1960s in Strasbourg, where Maurice studied geology and Katja physics and geochemistry. Both shared an enthusiasm for volcanoes. After graduation they founded a centre of volcanology and got married.

Maurice and Katja were very different both in appearance and character. He was big and strong, with an acute sense of humour, which he expressed through loud bursts of laughter. She was delicate, graceful, and much more reserved than her husband. But they were inseparable and very determined in their actions. 'I am the whale and Katja is the pilot fish,' Maurice used to tell their friends.

They worked around the clock, ready to jump on a plane anytime an eruption was announced or predicted in some remote corner of the world. Upon return, they edited the movies and wrote books filled with hundreds of photographs. Several of the twenty volumes they published in the next two decades received international awards.

In 1991 when Mount Pinatubo erupted, Maurice and Katja's footage of various eruptions as well as the lahar flow at Nevado del Ruiz convinced the Philippine president, Corazon Aquino, that an evacuation was necessary. Without knowing it, the Kraffts saved thousands of lives. But they were not so lucky. In June of the same year they went to Japan to shoot the eruption of Mount Unzen when a glowing cloud of hot gas, ash, and rocks, called pyroclastic flow, washed the valley they were climbing, killing them together with volcanologist Harry Glicken and more than thirty Japanese photographers and journalists.

National Geographic produced a documentary about the Kraffts' life and work with interviews and scenes they had filmed in their expeditions. Ironically, Maurice is heard saying: 'I am never afraid because I have seen so many eruptions in 23 years that even if I die tomorrow, I don't care.' The next day, he was killed.

At the door of their house in Wattwiller, France, people left flowers when the news of their deaths reached them. A memorial note placed there ended with the words: 'Thank you Maurice and Katja, thank you for everything you have done for us.' The street where they lived is now called *Rue Katja et Maurice Krafft*.

The courageous work of these people has helped to better understand and predict volcanic eruptions. Unlike earthquakes and tsunamis, which don't put scientists in more danger than anyone else, volcanoes are riskier to study. Fortunately accidents like the one that killed the Kraftts and their colleagues are much less frequent than depicted in fictional movies.

The progress made since the birth of volcanology a century ago has been astounding. Demystifying the 'chimneys of Hell' and unravelling their secrets has helped with issuing general forecasts and, often, good predictions. In spite of tragic events, like the one that killed the Kraffts, the loss of human life due to volcanic activity has been low since the Mount Pelée event. This achievement would not have been possible without the growth of science and the defeat of ignorance.

4. GIANT WHIRLWINDS: HURRICANES, CYCLONES, AND TYPHOONS

I am more afraid of the West Indian Hurricane than of the entire Spanish Navy.

William McKinley

I have never experienced a giant whirlwind, and with a bit of luck I never will. But hurricanes often make news, and—like everybody who doesn't ignore the media—I know what they look like. Although their satellite images rival the beauty of spiral galaxies, an encounter with them on the ground is not an occasion to enjoy. The force of the wind and the flooding that might follow put such events in the realm of horror.

Indeed, hurricanes, cyclones, and typhoons—names given to the same weather phenomenon in different parts of the world—have killed more people over the past forty years than any other natural calamity. In 1970 alone, a tropical cyclone in East Pakistan was responsible for some half a million deaths. The cost of these disasters is enormous.

A tragedy the media covered worldwide hit the southeastern United States in the summer of 2005. Hurricane Katrina breeched the levees of New Orleans, part of which lies under sea level. The flood covered 80 per cent of the city with up to six metres of water. Most residents fled, but about one fifth of the population could not be evacuated in time. Many of those stranded found refuge in tall buildings. The poorly coordinated rescue operations took many days, and more than 1,500 people lost their lives.

The television broadcasts of this catastrophe moved the entire world. Unlike the 2004 Indonesian tsunami, which hit all of a sudden, Katrina besieged the city for hours. Had the levees resisted the storm, the flooding would have been averted.

The sad side of the story is that several experts had warned about the poor state of the levees long before this tragedy happened. Unfortunately those responsible for reinforcing the structures ignored the problem. Still, the timely prediction of Katrina's turn towards New Orleans led to the evacuation of most residents. Had that forecast missed the point, many more people would likely have died.

The Great East Pakistan Cyclone

On 1 August 1971, 40,000 spectators gathered at Madison Square Garden in New York to attend *The Concert for Bangladesh*, an event that would linger in the minds of music lovers for years to come. Among the performers were Bob Dylan, Ringo Starr, Eric Clapton, Leon Russell, and Billy Preston. The main organizer was George Harrison, who had agreed to invest his energy in a good cause. The funds the concert raised went to UNESCO for the refugees of Bangladesh.

In November 1970, East Pakistan had been hit by a powerful cyclone, which killed almost half a million people.

The government, which resided in West Pakistan, did next to nothing for the victims. This disaster happened in the context of political uproar at the time of a strong separatist movement in the east. Catalysed by the cyclone disaster, the events developed into a bloody civil war. East Pakistan declared independence in 1971 and changed its name to Bangladesh. The New York concert was part of a common international effort to help the innocent victims of the conflict.

This was not the first time that a giant whirlwind changed the course of history. As early as 1274, a typhoon stopped the invasion of Japan by Kublai Khan (grandson of the famous Genghis Khan), who lost more than 13,000 men in the disaster. In 1281 he attacked the islands again, this time with a larger armada. But nature's fury defeated him a second time. Kublai's losses were so big that he never attempted a third war. Since then the Japanese came to think of typhoons as *kamikaze*, or protective 'divine winds'.

In spite of many storm tragedies in tropical and adjacent regions, no single whirlwind killed more people than the Great Cyclone of 1970. It started in Malaysia with a westward-moving depression, which came over the warm waters of the Bay of Bengal and developed into a tropical storm. The wind intensified rapidly, turning into a severe cyclone. A few days later it hit the mouth of the Ganges River as the ocean was at high tide. Winds reached 230 kilometres per hour, and the storm surge exceeded six metres.

Meteorologists saw the disaster coming, but there was no network in place to warn the population. The coastal plains and the numerous islands of the Ganges delta, which account for almost a quarter of East Pakistan's area, were flooded during the night of 13 November. Many people were sleeping when the water came. The few who climbed tall trees to safety had to fight the snakes that had escaped the deluge.

With nearly 46,000 out of 77,000 fishermen dead, 65 per cent of the fishing industry vanished in one night. This was a big blow for a country whose main source of nourishment is seafood. Other areas of the economy were also hard hit. The social, economical, and political consequences of the disaster were devastating.

A swampy area as low and large as southern Bangladesh is impossible to evacuate on short notice. So to avoid future disasters, the new government decided to build tall concrete shelters able to host up to 1,000 people each. But although financed by the World Bank, the project provided less than 300 buildings before 1985, when a strong tropical storm killed 10,000 people. The addition of eighty-six shelters during the following years could not prevent 138,000 deaths from a cyclone in 1991.

Soon after this tragedy, the number of shelters rose to more than 2,000, thanks to the help of non-governmental organizations. A million people took refuge in them during the powerful storm of May 1997. Fewer than 100 people died. Though any loss of life is deplorable, the difference between the number of victims before and after 1997 is staggering, and shows how investment in the right infrastructure not only saves lives but also prevents economic and social disasters.

Inside the Vortex

Hurricanes didn't concern the scientific community until about two centuries ago. Only remnants of hurricanes have reached

Europe in the past, so when the physics of the atmosphere began to be understood, the early experts, who lived on the old continent, were unaware of the devastation these giant vortices produced. No wonder that the first hurricane researcher was an American amateur meteorologist named William Redfield.

In 1821, he studied the damage inflicted by a strong windstorm, which had moved northward through the states of the mid-Atlantic and New England. He was surprised to see that the toppled trees pointed east along the coast and west in the interior. This effect could be produced by a giant whirlwind, a phenomenon similar to a tornado, but of much larger scale. William Reid, a British engineer who examined a hurricane disaster area in Barbados, reinforced this idea ten years later.

Redfield and Reid thought that hurricanes act only in the lowest one or two kilometres above the Earth because hills and small mountains appeared to break them apart. But Benito Viñes, a priest and meteorologist who studied the phenomenon in Cuba decades later, disagreed with his predecessors. He pointed out that hurricanes produce clouds consisting of ice crystals, which occur only in the cold layers of the atmosphere. Later findings proved Viñes right. Hurricanes can reach heights of up to eighteen kilometres.

Research in this direction was slow in the following years but grew significantly during the second half of the twentieth century, when aviation, radar, and satellite science became established. Besides photographic sights that helped atmospheric physicists understand how hurricanes develop and move, airplanes allowed experts to travel through the eye and the walls of the vortex, take measurements, and decipher the secrets of this weather phenomenon.

The hurricane's core has impressed scientists and poets alike. As satellite clips show, the eyewall has the highest rotation speed of the entire vortex. Indeed, the strongest winds and heaviest rains are registered in its neighbourhood. But the eye itself, which resembles a large tube, rotates slowly with gradually decreasing speeds towards the centre. The air, however, moves downwards inside the eye and upwards near the eyewall.

Those who have flown through the heart of a whirlwind recall unforgettable images. Kerry Emanuel, a professor of atmospheric science at the Massachusetts Institute of Technology, recounts his experience in the following words: 'No mere photograph can do justice to the sensation of being inside the eye of a hurricane. Imagine a Roman coliseum 20 miles wide and 10 miles high, with a cascade of ice crystals falling along the blinding white walls.' Such descriptions obscure the borderline between reality and fairytale.

Images alone, however, are not enough to understand the dynamics of a hurricane. Temperature, humidity, and pressure also play a crucial role. Radar signals are sent towards the hurricane, and the radiation scattered back by the various components of the vortex is captured and measured. Hailstones, ice crystals, snowflakes, and raindrops produce strong radar echoes, which are used to predict the drift of the system.

Data-collection techniques evolve and become more precise. Wind speed, for instance, is measured from an aircraft whose velocity must be known. In the past, the evaluation of the airplane's velocity was approximate, but the introduction of the Global Positioning System (GPS) allows excellent estimates these days.

Data collection, however, is only a first step towards predicting how giant whirlwinds move. The next stage is to understand
the physics of the phenomenon. Then mathematics takes
charge, from building a suitable model to solving the corresponding equations and obtaining the desired solution. So let us
now see how hurricanes form, grow, lose power, and vanish
from sight.

A Tropical Engine

Many of my fellow countrymen dream of retiring in the tropics.
Others would like to spend only the cold months there. Canadians who winter in Florida and other southern states, for
instance, are nicknamed 'snowbirds'. These people enjoy the
tropical climate between December and May, but avoid its
dangers during the other months of the year. Indeed, the
June–November period is known in the Caribbean as 'hurricane
season'.

The main reason hurricanes form is the greenhouse effect—
the phenomenon that raises the room temperature when the sun
shines, the windows are shut, and the curtains open. To some
extent tropical regions behave like the sunlit side of a house, while
areas less exposed to sunrays are more like Siberia or Antarctica.

Two sources heat the Earth: the Sun and the atmosphere,
which is also heated by the Sun. Calculations show that the
Earth would be much warmer than it is, if radiative heating
acted alone. But two phenomena cool the Earth's surface:
convection—the vertical currents that take hot air up and cold
air down—and water evaporation. We experience the action

of the latter when feeling cold as we emerge from the sea. Like us, the Earth releases energy to evaporate water into the atmosphere.

Once released, the heat warms the air with the help of convection currents. The water vapours form clouds, which stimulate the greenhouse effect. The atmosphere absorbs more heat from the sun and transfers energy to the Earth. Then the Earth evaporates more water, which thickens the cloud layer, and so on. This mechanism becomes highly effective during the warmer months. In marine tropical regions, the greenhouse effect combined with convection and evaporation sets the conditions for powerful storms.

In the 1820s, the French military engineer and physicist Sadi Carnot was the first who understood the principles underlying the hurricane engine. Though he died young, at 36, of cholera, Carnot made huge contributions to thermodynamics. The four-step process that maximizes the conversion from heat to mechanical energy is known as the Carnot cycle. He also derived the second law of thermodynamics, which explains why 100-percent-efficient heat engines cannot exist.

In his early youth, Carnot was fascinated with the steam engine James Watt had invented in 1784, and which marks a crucial moment in history: the start of the Industrial Revolution. Carnot wanted to grasp the physics behind Watt's machine. He realized that heat plays the most important role not only for making engines work but also in many natural phenomena. At the age of 26 he published a slim book in which he wrote:

It is to heat we should attribute the great movements that appear to us on earth; to it are due the agitations of the atmosphere, the rising of clouds, the fall of rain, the water currents that furrow the surface of the globe and of which man has managed to turn a small part to his use; and finally, earthquakes and volcanic eruptions are also caused by heat.

Carnot's engine works as follows. A gas is trapped in a cylinder endowed with a piston (see Figure 4.1). At step 1, the heated gas pushes the piston up. Consequently, the pressure inside the cylinder goes down. The temperature level, which should drop, is maintained with a heat source that is turned off at step 2. Then the piston continues to rise, while pressure and temperature decrease. Step 3 reverses step 1: the gas cools, the piston falls, and the gas pressure rises. Meanwhile, a coolant keeps the temperature constant. After the cold source is removed at step 4, the gas gets compressed until it reaches the initial temperature. Then the cycle starts again.

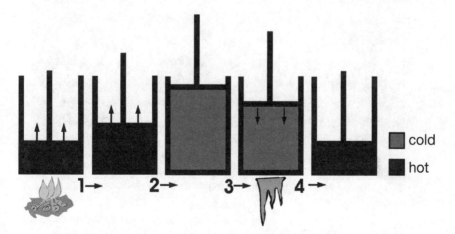

Figure 4.1 The four steps of Carnot's ideal heat engine.

Although car engines differ quite a bit from the above design, the hurricane resembles Carnot's scenario in a first approximation. The gas in this case is a combination of moist air, water drops, and ice crystals. It travels up through the eye of the hurricane and is expelled into the environment. But some new gas returns to the eye's bottom to continue a cycle similar to the one Carnot described. So though the gas is renewed, it behaves as if trapped along the path A–B–C–D (see Figure 4.2).

The eye of the hurricane acts like a vacuum cleaner, absorbing the air from the water surface. At step 1 (leg A–B), evaporation heats the gas. During step 2 (leg B–C), the gas is absorbed

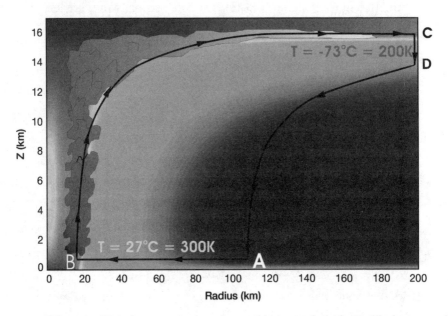

Figure 4.2 How the moist air, water drops, and ice crystal circulate inside the hurricane.

through the eye and slides along the surface of the hurricane. Pressure decreases but total heat remains unchanged. At C, the temperature goes down to −70 degrees Celsius. While the air descends (leg C–D), the temperature is maintained. At step 4 (leg D–A), the air sinks to the water surface. Heat loss and gain are balanced, so little heat is gained or lost in this leg.

The Carnot model allows an estimate of the maximum speed the wind can reach in various parts of the world. In general, those speeds decrease when moving away from the tropics, but local variations show up due to the ocean currents. For instance, the absolute top speed of 360 kilometres per hour could occur in most of the tropical Pacific Ocean, especially north and north-west of Australia, along the west coast of Mexico, as well as in the Indian Ocean and east of Africa. Such strong winds are less likely in the Caribbean, only perhaps around Cuba and the eastern region of the Gulf of Mexico.

But speeds between 210 and 270 kilometres per hour might show up in any tropical region, except west of Peru and northern Chile, which are as lucky as the temperate and cold parts of the globe. Should they form in those regions of the planet, hurricanes would not exceed 100 kilometres per hour. But they don't occur just anywhere. For various reasons, some areas are more prone than others.

The Chain of Creation

A colleague asked me recently why hurricanes originate only in certain parts of the globe and not everywhere. In particular he wanted to know why he often hears about hurricanes in the

Caribbean but rarely on the coast of Brazil. I took him to my office and showed him the map in Figure 4.3, which plots all the points of genesis for a period of three decades. He noticed three main belts: two in the northern and one in the southern hemisphere, and wanted to understand why this pattern shows up.

To explain, I started with some general comments. Experts, I told him, don't wonder why hurricanes occur, but why they form so rarely. Unlike tropical thunderstorms, which grow spontaneously, hurricanes need a push similar to the one your car's engine gets from the small electric motor attached to it. Without this initial impulse, hurricanes fail to develop, and end up as mere tropical rains.

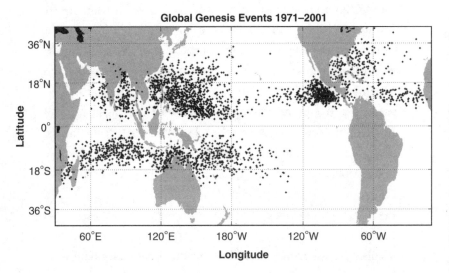

Figure 4.3 The original points where hurricanes, cyclones, and typhoons were generated between 1971 and 2001.

Hurricanes usually form in the tropical regions but stay away from the equator. Still, sometimes they come close to it. The record belongs to Typhoon Vamei, which originated at 1.5 degrees northern latitude in December 2001. The reason for this distribution is that whirlwinds develop in an atmosphere that has the tendency to spin. Since Earth rotates around an axis through the poles, the atmosphere (assumed at rest) has no spin at the equator but some spin everywhere else. Should the poles be as warm and humid as the tropical regions, giant vortices would be most likely there.

Humidity plays a key role in triggering tropical storms. While the air in the tropics is humid when rising, it dries when it sinks back to earth. Giant whirlwinds can form only if there is enough humidity in mid atmosphere. But this factor is not sufficient either. Winds that change speed or direction with altitude also inhibit hurricane formation because they bring dry air into the system.

Finally, hurricanes cannot occur if the basic heat conditions that trigger the Carnot cycle are missing. This explains why South America's west coast, for instance, is free of genesis points even though part of Peru and northern Chile lie in the tropics. In those places, the ocean's temperature is not elevated enough to fuel the development of hurricanes.

My colleague left my office happy with this explanation. Of course, I could have added more, such as the fact that not all the details of the formation process are clear today, and experts are unable to tell in advance whether a cluster of storms will start rotating. Therefore the attempts to predict the path of a hurricane begin only after the vortex is fully formed. Then, however, the big challenge occurs: to understand how hurricanes move.

Gone with the Wind

The tip of a spinning toy top rarely stands still on the table; it moves along an unpredictable path, performing loops, changing direction, speeding up or slowing down. This drift is due to various factors, such as how the top is launched, the uneven state of the table, or air movements in the room. Hurricanes are at least as sensitive to external perturbations as toy tops, so their wandering is difficult to understand.

The atmosphere is never completely still. Air moves at various speeds in different directions. In a first approximation, hurricanes go with the surrounding wind. But the vortex is not like an air balloon at the whim of the currents. Due to its rotation and large size, it has an impact on the motion of the flow.

If the power of the vortex is small, the direction of the background wind might offer a rough prediction of the hurricane's future motion. In some cases this method may work, for instance when the airflow is uniform and smooth. But this happens rarely. In general, it is either turbulent or has points in which directions change.

An example that illustrates the latter aspect appears in Figure 4.4. Between two high-pressure systems there is a *saddle point*. Assume that northern winds blow eastwards, and southern winds, westwards. If a southern (westward-moving) vortex comes close to the saddle point, it might keep its direction or could get caught in the northern flow, which would make it turn east. Predicting the future motion of a vortex near a saddle point is therefore impossible by looking at the background flow alone.

But even if the current is smooth, plain meteorological predictions are of little use because the vortex changes the

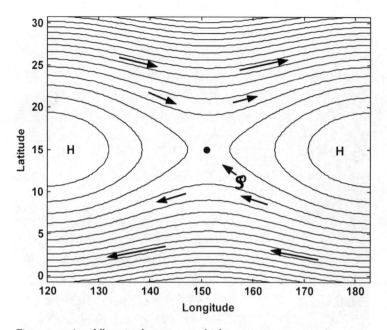

Figure 4.4 A saddle point between two high-pressure systems.

surrounding flow. The atmosphere tends to preserve the quantity of spin around the vertical axis. As a result, additional rotation occurs to compensate for the twist of the cyclone. This rotation pushes northern-hemisphere hurricanes northwest and southern-hemisphere hurricanes southwest by speeds of up to eight kilometres an hour. If not taken into consideration, this effect can lead to prediction errors of almost 200 kilometres a day. Such errors are large enough to make a three-day forecast useless.

The background flow, the structure of the vortex, and the compensation drift are the three components that determine the path of a hurricane. This combination can lead to bizarre tracks, like the ones of typhoons Ione and Kristen, which got entangled

Figure 4.5 Typhoons Ione and Kristen of 1974.

in a dance in 1974 (see Figure 4.5), or Hurricane Elena, whose sudden turn in 1985 spared Florida but hit Mississippi and Louisiana (see Figure 4.6). Though such phenomena may look exotic, they are no fun for those who experience them first hand.

Elena's Whims

Hurricane Elena rests in the memory of many Americans as a whimsical natural disaster which cost the country 1.3 billion

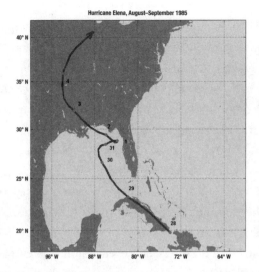

Figure 4.6 The path of hurricane Elena in 1985 from 28 August to 4 September.

dollars and forced the evacuation of nearly one million people from Florida, Alabama, Louisiana, and Mississippi. At the time, it was the largest peacetime evacuation in the history of the United States.

It all started in August 1985 near the Cape Verde Islands, off Africa's west coast, when a cloud cluster developed into a storm, crossed the Atlantic Ocean, passed along northern Cuba, and on the night of the 28th entered the Gulf of Mexico as a hurricane. With winds of more than 160 kilometres per hour, this giant whirlwind headed northwest towards Mississippi and Louisiana, where evacuations began.

But on 30 August, when about 300 kilometres from New Orleans, the hurricane slowed down and turned east. Florida's residents began to evacuate, while people from the previously threatened states returned home. Labour Day tourists also came to spend their short holidays in southern Mississippi and Louisiana.

Elena, however, changed course a second time. On 1 September it turned back when only eighty kilometres from Florida's coast and headed northwest to its original destination. Those who had just returned home had to flee again.

Labour Day 1985 is a sad memory for the residents of Biloxi, Mississippi, where Elena's eye came ashore. Though the population had been evacuated, the devastation was heartbreaking. Many other communities along the coast suffered heavy damage. Florida didn't escape completely, but its losses were light. Some children in the Sunshine State even found this event exciting, as did Ashley Wroten, who was growing up in Clearwater. Her description is unlike what most people would have written:

My dad, a pilot, was away on a trip—lucky for him but unlucky for my mother, who was stuck trying to evacuate with a young child. We packed our bags and headed off to a nearby elementary school that was serving as a shelter for evacuees. We set up camp along with thousands of other folks. I made many friends by wandering around and talking to complete strangers. I guess little kids lack the perspective and understanding about this type of situation, because I was in heaven. So many new people to talk to! The following day we were able to return home. There was, of course, lots of debris, but our home survived, and the adventure thus ended.

Ashley Wroten, like most people in Florida, had been lucky. The background wind coming from southeast had initially driven the hurricane northwest. But a high-altitude low-pressure trough moving across the mainland pushed westerly winds more to the south with the effect of changing the hurricane's path eastwards. As the trough vanished in the Atlantic, Elena drifted slowly for a few hours until the original wind resumed and guided her northwest again.

This example is typical of how external factors determine a hurricane's path. To capture such effects in their predictions, the experts use models based on the differential equations that describe atmospheric motion. But to solve these equations, they need the current conditions of the hurricane, which must be collected from inside it.

Hurricane Chasers

Hurricane movement would be impossible to forecast these days without specialized aircraft, which fight strong winds and turbulence to gather the necessary data. But isn't it dangerous to fly such missions? Nobody knew the answer to this question before the first attempt was made.

Until the early years of World War II, airplane pilots flew their machines visually with a map and a compass. This undertaking was risky. The sense of up and down is easy to lose in the clouds. To make flights more secure, additional instruments were added on board so planes could fly even in bad weather. But nobody dared to confront giant whirlwinds until two American officers had the courage to try it.

In late July 1943, a tropical cyclone originating in the Gulf of Mexico threatened southern Texas. Weather forecasts were a secret matter during the war, and the Air Corps command at the Bryan Army Air Field, 200 kilometres from the shore, had learned about the danger through military channels. Fearing damage to their AT-6 Texan Trainer airplanes, they ordered them evacuated farther north.

The British pilots who happened to be assigned in Bryan preferred the American fighters to the frail Texan two-seaters. But since the order to fly the AT-6 had to be obeyed, the Brits responded by making fun of them. The local pilots felt insulted. At some point, Major Joseph Duckworth, an American lead instructor, had enough.

'Any airplane can fly in any kind of weather,' he told the Brits. 'The only thing that matters is the skill of the pilot to use the instruments on board.'

'Then prove it!' a British pilot shot back.

'All right!' Duckworth said, and offered a bet, which was accepted on the spot.

Duckworth needed a navigator, but the only one in the cafeteria where the discussion was taking place was Lieutenant Ralph O'Hair, who seemed reluctant to go.

'There was no one in the Air Force who could have ordered me to do that,' O'Hair recalled later. 'I love life too much.'

But he respected Duckworth, who was by far the most skilled pilot at the base. If Duckworth said he'd do it, then it could be done. This thought convinced O'Hair to accept.

The two men decided not to seek permission from Headquarters. They knew they might not get it. Ignoring the thought of what would happen if their plane's only engine failed, the officers took off around noon. They had no idea that, through their act, the day of 27 July 1943 would become a landmark in the history of aviation.

Confronting winds of up to 160 kilometres per hour, they approached the vortex at medium altitude. O'Hair felt he was 'being tossed about like a stick in a dog's mouth'. The deeper they went into the hurricane, the darker and rainier it became. They heard the torrential water splashing against the airplane. Then, all of a sudden, brightness filled the sky. A curtain of clouds surrounded them. They had reached the hurricane's eye, which was shaped like a leaning cone, about fifteen kilometres across. The relief O'Hair felt was immeasurable.

After calling the Houston weather station to give their coordinates, they circled the eye a few times and headed back to Bryan. The return flight was easier because they had conquered their fear and felt more relaxed. Flying into a hurricane's claws didn't scare them anymore.

In the meantime everybody at the base had learned about Duckworth and O'Hair's adventure. When the AT-6 landed, Lieutenant William Jones-Burdick, the weather officer, asked Duckworth to take him back to the hurricane's eye for some visual observations. O'Hair descended, Jones-Burdick jumped in, and Duckworth took off again towards the vortex. Less than two hours later they were back. The first reconnaissance mission had been successful.

The excitement at Bryan was so high that several other pilots flew without permission into the hurricane, but using B-25 twin-engine airplanes instead of Texan Trainers. The collective fear of strong winds had been overcome. Systematic surveillance operations could now begin.

The success of Duckworth and O'Hair did not imply that hurricane missions were safe. In the early days of weather surveillance, tens of crew lost their lives on such occasions.

Low-altitude flights were especially dangerous, because the air is more turbulent than at higher altitudes, and a simple mistake or a sharp gust of wind can throw the plane into the sea. In the late 1950s, aviation engineers designed better instruments known as radars, which collected the data from higher altitudes by sending pulses of electromagnetic radiation towards the sea surface. The subsequent invention of Doppler radars improved the measurements and allowed planes to stay away from danger.

Still, not all the missions that followed were trouble-free. A near disaster happened in September 1989, when an NOAA P-3 four-engine aircraft flew into the core of Hurricane Hugo over the North Atlantic. As the plane crossed the wall cloud into the eye, wind speed dropped from 280 to 30 kilometres an hour, while a defective fuel valve produced an engine fire. The pilots cut the power and dumped some fuel to stop the plane's free fall. This risky operation paid off. The aircraft stabilized, climbed, and landed safely in the end.

Due to the progress of aerospace engineering, flights into tropical cyclones have become secure. Passenger planes on routine flights often experience rougher turbulence than that met by surveillance pilots. The modern airplanes are safer than the two-engine machines used in the past, and most flights into hurricane cores prove uneventful.

The future of reconnaissance missions appears to be bright. Today's experts test robotic aircraft, design better data-collection instruments, and try more sensitive methods of measurement. When these developments reach maturity, unmanned airplanes will help gather highly accurate information.

Making Predictions

How do scientists predict the future path of giant whirlwinds, and how precise are their forecasts? To answer these questions, let us take a look at what scientists did in the past to understand the motion of hurricanes and compare the evolution of their methods with the current state of the field.

Hurricane prediction belongs to the larger effort of weather forecasting, whose roots get lost in the history of humankind. People have always tried to foresee tomorrow's weather. Their 'science' was based on simple observations, which have been passed by word of mouth from each generation to the next. But tropical cyclones enjoyed a special status. Since they were deemed divine and beyond human understanding, the local population did not question them. Therefore most prediction attempts we know about are from after the discovery of the Americas.

In his voyages to the West Indies, Christopher Columbus learned from the native people to recognize the signs that indicate the approach of powerful storms (wind speed, sea swell, and typical cloud formations) and succeeded in predicting the first hurricane he would ever encounter. For more than three centuries after Columbus, pattern recognition was the only way to forecast these extreme winds. Unfortunately such warnings were often too short, so they were of little help in preventing disasters.

Things changed only in the second half of the nineteenth century, when Father Benito Viñes, the Spanish Jesuit priest mentioned earlier in connection with the height of giant whirlwinds, arrived in Cuba and became director of the meteorological observatory at the Royal College of Belen in Havana. When he witnessed the devastation hurricanes produced on the island, Viñes decided to research them. He organized daily weather observations, which he later extended across the Caribbean, and collected information from old newspapers to enrich the records the observatory already had. He also checked the sites devastated in the past to find clues about wind speed and direction.

His efforts paid off. Father Viñes recognized that the motion of clouds could be used to locate the centre of a storm and make a short prediction of a hurricane's path. He further established certain local wind and current patterns, which laid the foundation for weather forecasts in the Caribbean.

On 11 September 1875, Father Viñes predicted the first hurricane two days before it reached Cuba. The path of devastation he foresaw was accurate enough to give the islanders time to take cover and protect their belongings. Other predictions followed. Soon Father Viñes became highly respected both in the scientific world and in the Caribbean community, which heeded all his warnings.

His prestige reached such heights that steamship companies offered him free trips to help him with his work. The captains had orders to make unscheduled stops if he had to visit an island. But in spite of all the attention given to him, the hurricane priest—as he became known—remained a modest man. When he died, in 1893, the newspaper *La Lucha* published a special edition entitled 'The Last Hours and Death of Father Benito Viñes'.

Research in the Caribbean and other parts of the world continued in the following decades, and the development of atmospheric physics allowed a better understanding of hurricane formation. But until the mid 1930s, all measurements and observations were made at the Earth's surface. Things changed when the first weather balloons were launched at high altitudes, thus significantly improving the acquisition of information. After World War II, aircraft surveillance took over data collection in hurricanes.

Before the mid twentieth century, the only difference between the scientific method and folklore was that of scale. The modern experts had a single advantage over Columbus: a larger and more accurate pool of data, but the margin of error of their predictions was still large. All that changed in 1950, when Jule Charney, Ragner Fjørtoft, and John von Neumann from the Institute for Advanced Study in Princeton employed an early electronic computer to produce the first automated numerical weather forecast. Though modest, their approach signalled a fundamental change of direction: they used a mathematical model for the first time.

Shy attempts of this kind had existed in the past. The Norwegian physicist Vilhelm Bjerknes suggested in 1904 that weather predictions could be made with the help of the differential equations that describe the motion of the atmosphere. The British scientist Lewis Fry Richardson took this idea seriously. In 1922 he thought of gathering thousands of human calculators and assembling their computations into a forecast, but ended up doing the calculations all by himself. The wrong projection he obtained after several months of work put an end to his enthusiasm.

The invention of the digital computer allowed a reexamination of these ideas. John von Neumann gave hope to all those interested in predictions. His first endeavours to use computer power in conjunction with numerical methods showed excellent results in calculating the future positions of the planets. Von Neumann's pioneering work with Charney and Fjørtoft raised the expectations of meteorologists too.

But in 1961 the hopes nurtured for almost a decade were shaken. Edward Lorenz, who was a meteorologist at the Massachusetts Institute of Technology, rediscovered the chaos phenomenon while using a simple model of the atmosphere. The differential equations he tried to solve numerically manifested high instabilities. Slight variations of the initial data led to large changes in the long run. This meant that precise weather predictions could never be given well in advance.

Indeed, weather forecasts are now limited to a maximum of about two weeks. This threshold is unlikely to be beaten in the future, even if computers improve their speed, capacity, and precision. Luckily, a warning of two or three days is usually good enough to evacuate the regions a hurricane would affect. And though the experts can make timely predictions in most cases, there are times when they cannot.

Figure 4.4, for instance, points at such an example. If the early calculations show that the hurricane is heading towards what the experts call a saddle, it is hard to predict whether the path beyond that point will turn in one direction or the other. But once the turn is made, a renewed forecast becomes reliable until another trough or ridge impacts the hurricane's course.

Though the few-day weather-forecast limit is unlikely to expand in the future, there are ways to make more accurate predictions. For this purpose, scientists are working in several directions: to obtain better measurements, improve the mathematical models of the atmosphere, come up with more precise numerical algorithms, and find ways to initialize the equations faster and more efficiently.

The last aspect is specific to weather. To predict celestial motions, for instance, astronomers must feed their numerical algorithms with the present positions and velocities of the planets, whose number is very small. So punching the data into the computer is fast and easy. But to make a realistic weather forecast, meteorologists have to start from the current atmospheric conditions and use thousands of local measurements. Therefore the process of initializing the programme is much more complex than in celestial mechanics.

To overcome these practical difficulties, hurricane experts use a twelve-hour-old forecast to bring the computer simulation close to the current conditions. Then they compare the state of the model with the new measurements and modify their first guess. This approach may appear clear enough not to cause headaches, but its implementation becomes far from simple when the accuracy of the forecast is the main goal.

Hurricane prediction is an active research field these days. There are many forecast centres spread over the tropical regions. They issue reports every six hours with long- and short-term projections. Once they spot a hurricane, they describe its path for the next two or three days, often succeeding in forecasting the future coordinates of the eye within a few tens of kilometres. The six-hour predictions are, of course, more precise, but they provide little help if large evacuations are needed. What experts are still struggling with is forecasting the intensity of hurricanes. In spite of many efforts, this aspect of their research has not produced satisfactory results.

So though lacking a magic formula to describe the path of a tropical cyclone days in advance, the experts have a good grasp on predictions. Moreover, the progress of science and technology helps them improve their forecasting techniques. But they also know there are limits to what can be done. And this leads to a dilemma. Should the authorities always issue an evacuation order when a hurricane forms, even if the chances that it will hit a certain spot are small but non-negligible?

Sounding the Alarm

Imagine the following scenario. A strong hurricane originating southeast of Florida accelerates towards New York. The Hudson River is swollen due to a rainy period. The computers indicate that the hurricane's eye will hit Manhattan at maximum tide. The water surge could reach seven metres. But some last-minute measurements show the existence of a weak atmospheric trough, which might deviate the hurricane away from New York.

An evacuation of several million people at such short notice would not only cost billions of dollars, but could also lead to hundreds of deaths due to accidents, fights, health problems, and panic. Looting would be hard to prevent. Offering shelter, fresh water, and food to these people—even if only for a few days—poses hard logistic problems. All this pain would be worthless if the hurricane didn't hit New York. But if it did, the death toll could be immense. Should the authorities evacuate the city? I guess that nobody would want to make this decision. Still, someone must have the final say.

The trouble with this scenario is that it's not impossible. High levels of the tide, a swollen Hudson River, and strong winds have happened in that region in the past, but not all at once, so a situation similar to the one described above could happen someday. It was New Orleans that was hit in 2005. It might be New York next time. Nobody understands this danger better than the experts in hurricane prediction. Therefore several of them issued a warning on 25 July 2006. They started their appeal as follows:

As the Atlantic hurricane season gets underway, the possible influence of climate change on hurricane activity is receiving renewed attention. While the debate on this issue is of considerable scientific and societal interest and concern, it should in no event detract from the main hurricane problem facing the United States: the ever-growing concentration of population and wealth in vulnerable coastal regions. These demographic trends are setting us up for rapidly increasing human and economic losses from hurricane disasters, especially in this era of heightened activity. Scores of scientists and engineers had warned of the threat to New Orleans long before climate change was seriously considered, and a Katrina-like storm or worse was (and is) inevitable even in a stable climate.

The message is clear: the only way to prevent hurricane tragedies is to stop the growth of cities in zones that, sooner or later, will be hit. Only small communities can be safely evacuated when a tropical cyclone strikes. So, in the long run, the dilemma of sounding the alarm if it's not clear whether a hurricane will hit a big city can be solved through preventive planning. Of course, this solution doesn't remove the problems of the New York scenario, which have no straightforward answer.

Since they are no experts on climate, the signatories of the 2006 appeal did not dwell on whether the recent hurricane activity warns of climate change. But those who study those problems claim to know what can happen, and have made their voices heard. Let us now see how they reached their conclusions.

5. MUTANT SEASONS: RAPID CLIMATE CHANGE

Our response to the threat of global warming will affect our personal well-being, the evolution of human society, indeed all life on our planet.

Spencer Weart

About an hour's drive from my native city in Romania is Copşa Mică, which was until recently the most polluted town in Europe. In the mid 1980s, when I often passed through that region by train, the place looked like hell on earth. Everything, from streets and houses to trees and hills, was black from smoke. Grass and flowers had long vanished, and people's life expectancy was far below the national average. The proximity of the Carbosin factory, the apartment blocks for its workers, and the cemetery added to the grotesque.

This place made me aware at an early age of how ruthless we can be with nature. So I was not surprised to learn that our lifestyle may lead to climate change. Nevertheless, the details of this process were not clear to me, mostly because the messages we hear are highly politicized. Therefore I decided to get better acquainted with the various points of view and try to distinguish between truth and fiction.

Recently, I witnessed an ongoing debate between two colleagues at the University of Victoria. Andrew Weaver, a climatologist with the School of Earth and Ocean Sciences, claimed that our carbon-dioxide emissions lead to global warming. Jeffrey Foss, a philosopher with an interest in the environment, disagreed. In his view, the climate is continuously changing, and there is no proof that we are responsible for the current trend.

I was thus confronted with two different points of view about global warming. Andrew had deeply researched a narrow domain, using mathematical models to make his predictions. Jeff had touched the surface of many fields, claiming that aspects he had found relevant for climate change were absent from the models used today. Andrew was a specialist, Jeff a generalist. Whom should I believe?

Under normal circumstances, I would have given Andrew all the credit. After all, who knows better than an expert? But climate change is not a standard branch of science. In his book, *The Weather Makers*, Tim Flannery warns that 'because concern about climate change is so new, and the issue is so multi-disciplinary, there are few true experts in the field'. Therefore I could not dismiss an approach that combines many angles and points of view before hearing the arguments.

It turned out that most of Jeff's objections didn't stand a careful scrutiny. He was correct when saying that the climate oscillates, but the changes don't seem to be internally chaotic, as most global-warming deniers claim. The variations in average temperature occur only because of some external impact.

Indeed, the average temperature of the Earth's atmosphere depends mostly on the heat received from the Sun. The laws of physics teach us that, under steady conditions, this temperature should stay about the same. But the conditions are not steady. Two kinds of factors influence them. The first ones, which are independent of the atmosphere's composition, make the intensity of the heat source vary for reasons such as the increase of solar luminosity or adjustments of the Earth's orbit around the Sun. The second ones change the amount of solar radiation the atmosphere absorbs. This variation can happen, say, during periods of intense volcanic activity, but it can also be ours in making. There is overwhelming evidence that, through greenhouse-gas emissions, we are responsible for the current trend.

These are the main conclusions I reached while immersed in the realm of good science, far away from politics. The following pages will get into details, surveying some basic achievements of climatology and explaining how predictions in this field are made.

A Shift in Opinion

Although the French mathematician and physicist Jean Joseph Fourier formulated the greenhouse effect as early as 1827, and the British scientist John Tyndall understood three decades later that the atmosphere controls the Earth's temperature, the common wisdom at the end of the nineteenth century was that the climate changes slowly, if at all. Some slight variations may exist, scientists thought, but they occur in thousands of years, and nobody can detect them in a lifetime. The ice age seemed to have been a local exception, whose causes were not clear. But then some evidence against this belief emerged.

It started with the Swedish scientists Svante Arrhenius, who tried to solve the riddle of the ice age. In 1896, he began to think of what could happen if the amount of carbon dioxide in the atmosphere varied, and concluded that an excess of this gas may account for future climate variations. For Arrhenius, however, this was a purely theoretical problem. He had no reasons to suspect that changes were in sight.

In 1911, the American astronomer Andrew Ellicott Douglass asked if there was any connection between the size of tree rings and sunspot activity. He thus invented dendrochronology, the science of tree-ring dating. One of his reports showed variations in temperature during the past few millennia, but his findings were dismissed as local at best. No scientist was ready to accept that Earth's climate could vary rapidly.

Some of those unconvinced that changes could be only local were curious enough to look deeper into why the climate had been cooler in the past. Among them was a Serb mathematician named Milutin Milankovitch, who came up with a global answer for this oscillation at about the same time Douglass made his discoveries.

Soon after he became a professor at the University of Belgrade, Milankovitch thought that celestial motions might be the reason for climate change. After all, summer and winter take place because the axis of our planet is tilted relative to the plane of its orbit. For six months a year, solar radiation heats one hemisphere more than the other. Could it be that the climate was determined by similar, but slower variations?

Between 1912 and 1914, Milankovitch published three papers in which he outlined an ingenious theory. Three factors influence the amount of radiation we receive from the Sun, he argued: the Earth's orbit, which varies in 100,000 years from almost a circle to a more elongated ellipse; the up-and-down tilt of the Earth's axis in the plane of the orbit in a 41,000-year period; and the orientation of the Earth's axis relative to the distant stars, due to our planet's top-like wobbling, with one rotation in 26,000 years.

Due to the start of World War I, nobody noticed Milankovitch's papers. Moreover, the Serb mathematician got no chance to promote his ideas because the occupying forces of the Austro-Hungarian regime interned him in Budapest. But he was allowed to work in the library of the Hungarian Academy of Sciences, where he continued his research and wrote a book, which appeared in print after the war.

Though Milankovitch explained the ice ages and gave some credibility to the idea of climate change, his theory predicted alternating periods of cold climate in the northern and southern hemisphere at intervals of about 10,000 years. This meant that the southern hemisphere was about to experience an ice age. As there was no physical evidence in this direction, the scientific world dismissed his results. The belief in a slow evolution of at most 1 degree Celsius per millennium was not shaken.

Some scientists, however, suspected a faster pace of change. One of them was Charles Ernest Pelham Brooks of the British Meteorological Office. In 1926, he published a book entitled *Climate Through the Ages*, which would become a classic. Brooks denied celestial motions any role in climate dynamics. He used a crude model to argue that the horizontal and vertical ocean currents may lead to large variations in global temperatures. A slight change in conditions, he claimed, could trigger a new climate regime. If during a warmer year the snowmelt exposed more ground, Earth would absorb more heat, warm the air, melt more snow, and change the climate in a few years. A reversed effect would produce a new ice age.

Most scientists met Brooks's theory with scepticism, and it would take them two decades to change their minds. In the meantime, the British scientist received recognition from religious groups, who used his conclusion to support the Earth's biblical age of about 6,000 years. Though not in Brooks's intention, this interpretation helped to spread his ideas, and soon some experts reconsidered his work.

One of them was W. J. Humphreys, a meteorologist with the US Weather Bureau. In an *Atlantic Magazine* article published in 1932, he warned about the possibility of disrupting the Gulf Stream. Should this happen, he said, the Old Continent would be rapidly gripped by an ice age. This idea was not new, but few experts had heard about it. Humphrey also suggested other scenarios, such as a series of major volcanic eruptions, which would diminish the sunlight and allow the build-up of ice.

At about the same time, scientists noticed that the climate of North America had grown warmer in the past decades, but attributed the change to some natural cycle, whose mechanism they didn't really grasp. There was, however, a man who didn't accept this explanation, the British engineer and amateur meteorologist Guy Stewart Callendar. In 1938, he argued that the rising global temperatures were linked with coal burning, thus pointing out the role carbon dioxide plays in climate change, as Arrhenius had suggested four decades earlier. Callendar supported his conclusion with data from around the world and a coherent physical theory of infrared absorption by greenhouse gases.

Again, the great majority of scientists ignored these results. The ocean-floor samples showed no disturbance of the layers deposited in tens of thousands of years, so the climate must have been stable. In fact, a combination of bad sampling and wishful thinking had affected their interpretations. As the techniques improved in the late 1930s, samples taken from Scandinavian lakes suggested a different story. Vegetation residues revealed that warm- and cold-climate species had lived there through the ages.

The 1949 invention of radiocarbon dating revolutionized climatology. Though the method gave large errors during its early stages of development, and many comparisons between ocean-floor samples didn't make much sense, some scientists grew convinced that Earth's climate had suffered rapid changes in the past. The temperature variations Andrew Ellicott Douglass had read in old tree rings didn't look absurd anymore.

An unexpected event, however, would give a bad name to those who had begun to see how fast the climate could change. A Russian medical doctor caught the public eye with a theory that would enrage the academic world.

The Frozen Mammoths

Worlds in Collision appeared in 1950, and sold millions of copies all over the globe. Its author was Immanuel Velikovsky, a 54-year-old Russian physician who lived in the United States. Before publication, *The New York Times*, *This Week*, *Herald Tribune*, *Pathfinder*, *Collier's*, *Vogue*, *Harper's*, *Newsweek*, and *Reader's Digest* deemed the book revolutionary. They hailed Velikovsky as a genius who had changed the way we understand history and the physical sciences.

The Russian doctor had cited more than a thousand sources in support of a catastrophic theory, which conflicted with the traditional timetables of the past. His revision of historical chronology was based on what he interpreted to be observations of a huge comet that had revolved for centuries on a stretched ellipse around the Sun, passing near us twice and halting the spin of Earth each time. After two close encounters with Mars, the comet cooled and became the planet Venus.

Scientists and scholars mocked Velikovsky's ideas. But the Russian complained publicly of how the academic establishment trashed his book before reading it. He cited many questions science was unable to answer, among them the uneven distribution of the ice caps during the latest ice age. His skilful rhetoric gave the impression he was winning the debate. Even if he never convinced any serious scientist about his theory, his editorial success prompted him to write sequels to his book.

In 1955, Velikovsky published *Earth in Upheaval*, in which he addressed climate change, a phenomenon he explained through his catastrophic scenario. Since average temperatures are almost constant over millennia, only the ramblings of his comet, whose approach had upset the direction of the Earth's axis, could account for the discovery of frozen mammoths with undigested food in their stomachs. Those animals had experienced an instant climate shift. The experts who had discovered rapid temperature drops in the past were proving him right. Or so he thought.

His arguments, however, are easy to refute. The mammoth literature reveals aspects Velikovsky omitted. The frozen specimens were mutilated and rotten, and their veins contained coagulated blood, which suggested death by asphyxiation. Endowed with woolly undercoats and fatty bodies, mammoths couldn't have died of cold. Also, out of about 50,000 specimens estimated to have populated the Arctic, only about forty were found. No other species shared this fate except for a few woolly rhinoceros.

There are many theories about the demise of the mammoths, but none of them is sound. A recent explanation by Richard Firestone, a nuclear chemist with the Lawrence Berkeley National Laboratory in California, argues that the animals died when an exploding star sent a wave of cosmic rays 13,000 years ago. Most scientists don't embrace this theory either. Nevertheless, whatever the reason for the mammoths' disappearance, it is clear that they didn't freeze due to a sudden change of climate.

In spite of his public success, Velikovsky gained a bad reputation in the eyes of scientists and scholars, who revealed faults and omissions at every step of his reasoning. Therefore anybody associated with him was viewed with suspicion. (Recently, I heard someone comparing global-warming deniers with Velikovsky— a total shift in public perception in just over fifty years. I don't know whether this analogy stands, but one difference between them is clear: unlike the Russian doctor, the deniers play a tune some corporations like.)

The Velikovsky episode tempered the enthusiasm of climate-change supporters. Luckily, the backlash didn't last long. The accumulated evidence began to point in a single direction.

Early Warnings

In the late 1950s, a team of scientists based in Chicago simulated the motion of the atmosphere by rotating a disk-like box containing a fluid. Instead of exhibiting a stable regime, as most climatologists expected, the fluid oscillated with quick transitions between two different states. Though far from reproducing the physical reality, this experiment seemed to bear the main features of climate dynamics. It was later revealed, however, that this experiment was more relevant to time scales of days or months, therefore better suited for weather rather than climate patterns.

Inspired by these results, Reid Bryson from the University of Wisconsin-Madison built in the mid 1960s an interdisciplinary team to look at climate from a novel point of view. He wanted to understand why the North American cultures of the Midwest went into decline in the thirteenth century AD. The historical evidence he found was astounding: the climate had shifted in less than three generations.

By the same time, the radiocarbon expert Hans Suess, a founding faculty member of the University of California in San Diego, announced the results of his study of deep ocean plankton shells embedded in cores of clay. His analysis showed many rapid climate changes in the past several hundred thousand years. This conclusion agreed with various reports coming from researchers who had looked at glacier moraines, deep ice cores, shells, pollen, and other elements indicative of average temperature variations. Climatologists all over the world were rapidly embracing the new theory of quick transition between climate regimes.

Now that this fact became clear, the main question to answer was why the change occurs. Milankovich's theory failed to explain all those shifts. It was already obvious that some catastrophic event, like a sequence of Krakatoa-like volcanic eruptions or the collision with an asteroid, could make the atmosphere less transparent, block a certain percentage of solar radiation, and trigger a new ice age. But the periods of climate change in the history of the planet coincided with catastrophic events only in few cases, so other external factors must have accounted for the variation.

The experts knew that water vapour, carbon dioxide, methane, ozone, and many other gases influence the temperature of the atmosphere, but not by how much. Therefore research focused on this issue. It turned out that water vapour is the most important factor, causing between 36 and 70 per cent of the greenhouse effect. But though the other gases have little direct influence, they affect air humidity.

The percentage of these gases varies in the atmosphere. Carbon dioxide, for instance, was between 260 and 280 parts per million (ppm) during the last ten millennia and is now 390 ppm. Nevertheless, some recent research shows that, in the past hundreds of millions of years, the amount of carbon dioxide had a minimum of 250 and a maximum of 2,000 ppm. Both the oceans and the land contain it, and its release in the air can happen naturally or through our activity.

The measurements of the American scientist Charles David Keeling, done in the late 1950s, showed an increase of the carbon-dioxide level in the atmosphere. As a result, more research was funded in this direction. The subsequent project led to what is known today as the 'Keeling curve', obtained at the Mauna Loa Observatory in Hawaii (see Figure 5.1).

The early results obtained at Mauna Loa convinced the experts to assess the impact of the carbon-dioxide increase on the global average temperature of the atmosphere. The first important

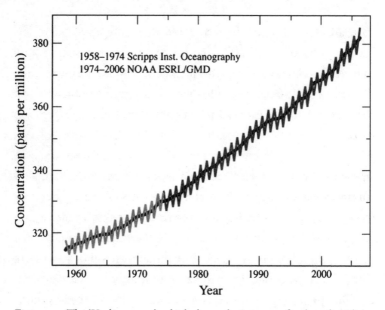

Figure 5.1 The 'Keeling curve', which shows the increase of carbon dioxide in the atmosphere between 1958 and 2002.

simple model, built in 1968 in Leningrad (now St Petersburg) through the work of Mikhail Budyko, predicted that, under the current rate of emissions, the ice of the Arctic Ocean would melt completely by the middle of the twenty-first century. Other researchers reached similar conclusions at about the same time.

This forecast was only the beginning of what would become a new research method in climatology. The experts understood that mathematical models implemented on powerful computers could produce forecasts. In the next few years, progress in this direction was remarkable, but the results were not pretty. The computations predicted an average temperature increase in the next decades, and the physical data agreed with this result. Our planet seemed to be experiencing some rapid climate change.

The Models

The early mathematical models of the atmosphere were simple. They used the classical equations that govern weather and climate physics, applied to an Earth partitioned into boxes 600 kilometres wide and with five vertical layers. Computations were done in each box. Some basic phenomena, like rain and the effect of radiation on carbon dioxide and temperature, appeared in the picture; others, such as clouds and air circulation, did not. The predictions, which ranged from a few months to a year, were far from the physical reality.

These results did not discourage the experts. They had better models in mind, but the existing computers were not powerful enough to process the required calculations. Nevertheless, hardware and software engineers assured them of a great future. Computers would grow both in size and capacity and would handle huge amounts of data. Some North American universities constructed large buildings in anticipation of tomorrow's super-sized machines.

Luckily only the latter prediction proved wrong. Computers became smaller and more powerful, and they changed our lives. Climatologists learned quickly how to make the best out of these achievements. They began to use either the new supercomputers or to connect many small machines into networks of high performance. Now they could add features to their models to make them more realistic.

Among their priorities were the reduction in size of the boxes that cover Earth and a more detailed evaluation in each box. The current computers can handle boxes ten to fifty kilometres wide with tens of vertical layers. Of course, this achievement is far from ideal, but there is little to be done before computer power increases many more times. Today's most sophisticated models need about three months to run a programme.

The new models contain clouds, most of which cool Earth during the day and trap the heat through the night, when they behave like a wool blanket. The percentage of the covered sky, however, was not the only feature to consider. Climatologists also modelled cloud structure. Indeed, the low, thick clouds we see on rainy days have a different effect on climate than the thin clouds formed in the higher atmosphere, so it was important to distinguish between them.

Another feature the experts modelled was the difference between land and water. Oceans take longer than land to heat up and cool down, and their currents have a crucial influence on climate. The Gulf Stream, for example, which brings warm water to Europe's west coast, is deemed responsible for keeping temperatures high there. In its absence, the old continent would probably have more severe winters than now.

Land could not be treated uniformly either. The contrast between light and dark regions had to be modelled too. Ice-covered areas, such as the Arctic, and deserts, like the Sahara, reflect most sunrays, while soil- and vegetation-covered regions absorb large amounts of heat from the Sun.

The balance between emitted and absorbed greenhouse gases was also a factor to consider. The oceans and the biosphere filter a lot of the carbon dioxide we produce now. This trend, however, will not continue forever, and it is difficult to assess the future emission–absorption ratio. Indeed, we cannot be sure about the exact number of people who will inhabit our planet fifty years from now, or how their lifestyle will impact emissions, and we cannot predict the extent of the biosphere and the capacity of the oceans to keep filtering greenhouse gases.

But these uncertainties are not the only issues difficult to model. The experts also have problems with assessing the effect of aerosols on climate change. Aerosols are what most of us call pollution: dust, ash, black carbon, sulphate, and other particles spread in the atmosphere by natural phenomena, such as forest fires and volcanic eruptions, and by our burning of fossil fuels. Aerosols reduce the transparency of the atmosphere and therefore cool the Earth. The long-term consequences of this effect are unclear.

Brown clouds have appeared in the past decades above cities. Recently, however, a huge haze began to span above parts of Asia and the Indian Ocean. In 1999, a team of 200 scientists researched this cloud and concluded that we contributed 75 per cent to its making. Black carbon seems to be its most damaging component, both for the biosphere and the climate.

The team measured that the reduction of sunlight below the cloud led to almost half the heat energy of what climate models predicted. The modelling complications increase because of an added effect: aerosols produce more cloud drops, which reflect solar radiation back into space, thus further cooling the planet. Moreover, black carbon inhibits rainfall in three ways. First, reduced sunlight leads to less evaporation; second, a cooler Earth produces a layer of cold air under a layer of warm air, an inversion that also decreases precipitation; and third, the haze slows down the summer monsoon.

Air pollution is not just a local problem. Wind spreads aerosols all over the world, no matter where they originate. Reducing pollution for health reasons will accelerate global warming if greenhouse gases are not equally reduced. The problem is that aerosols settle back to the ground much faster than gases. But obviously we cannot pollute the air at the current rate without damaging the biosphere. The example of Copşa Mică suggests what to expect if we ignore the problem.

The presence of aerosols has also hidden some of the warming our planet experiences. Oceans have stored the infrared radiation that penetrated the dimmed atmosphere, and it is not clear when they will release the accumulated heat; it could happen as soon as two or three decades from now. So the apparent cooling due to pollution is not only a factor very difficult to capture in mathematical models, but also one that might lead to a steeper rise in temperatures than we expect.

For reasons such as the presence of aerosols, models alone don't tell the whole story. Then how can we be sure that the planet is warming? What if there are still factors we don't know about, which will maybe lead to cooling in the near future or, hopefully, keep things as they are? To answer these questions, we must look at things from a new angle. Other disciplines reveal aspects we haven't yet seen.

Current Evidence

Mathematical models would be worth little if their forecasts disagreed with what we can measure. But when scientific prediction and the data obtained in the field match, it would be better to act before it's too late. The physical evidence leaves little doubt that our planet is warming.

The first indicator supporting this forecast is the rise of Earth's average temperature. The graph in Figure 5.2 depicts the oscillation of the 'global thermometer' between the years 1880 and 2000. During the second half of the twentieth century, the temperature increased by 0.6 degrees Celsius, with an abrupt surge in the last two decades. The new millennium shows no change of pattern, in agreement with the mathematical models, which predict a sharper increase in the years to come.

Not everybody agrees with these numbers. Some point out that most thermometers are in cities, which have grown warmer in the last decades. This remark is misleading. Cities do influence warming, but their contribution is very small. Measurements in the Arctic, where there is no urban population, show a

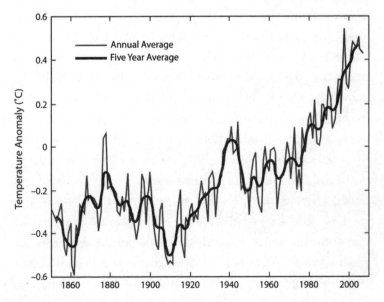

Figure 5.2 The variation of the global averaged temperature during the past 150 years.

larger variation in temperature than everywhere else. Still, some regions, such as parts of the southern oceans, have registered no change. One reason for this delayed response to warming is probably the one mentioned earlier: oceans take longer than land to emit the absorbed radiation.

Proof of temperature increase lies with the decline of snow for medium and high altitudes. In the northern hemisphere, for example, 10 per cent of this snow has melted since the late 1960s. Glaciers all over the world have also retreated. The most spectacular loss of ice took place on ice shelves. But we don't need to go that far to notice the change. Mountain lovers can witness this effect with their own eyes.

I saw it too, first time in the Canadian Rockies, while driving by the Athabasca Glacier in 1991 and again twelve years later. The difference between the pictures I took each time was staggering. The second opportunity occurred in the summer of 2004 at Grindelwald in the Swiss Alps, where a sequence of photos and a graph were posted outside a chalet built on a cliff above the glacier. The graph of the land's advance was not linear: rather, it looked like the one in Figure 5.2. But the melting tendency was obvious. In fact, the volume of glaciers in Switzerland has decreased by two thirds in the past century.

Again, global-warming deniers do not accept this evidence either, claiming that only convenient glaciers are chosen. But then how could we explain the rise of the oceans? Between 1961 and 2003, the global sea level went up at an average rate of 1.8 millimetres per year, with an accelerated rate in the last decade of the period.

The evidence for global warming doesn't stop here. Biologists have discovered behavioural changes in many species. Some plants that enjoy cooler temperatures have migrated up in altitude by several metres per decade. In temperate regions, gardens have shortened their dormant period. Every year, tens of bird species are laying their eggs between one and two weeks earlier than three decades ago. Moths, beetles, butterflies, and other insects have moved to higher altitudes, where they were not seen in the past.

These changes may seem insignificant, but not if we put things in perspective. Forests now growing in Canada, for instance, could be found 14,000 years ago only in northern Florida. Due to climate warming, they migrated north. At the extreme of this trend are extinctions. Proved to have died because of global warming are amphibian species, like the golden toad, that vanished from Central America in the late 1980s. The killer was a pathogenic fungus, which thrived under a slight rise in temperatures.

Some scientists suggest additional evidence of global warming. Typical examples are the surge of drought in some regions, with an increase in the number of large forest fires, as the ones that devastated part of Greece and California in 2007, rain and floods in others, and especially hurricane activity in the tropics. These aspects of the weather are popular with the media, but the link seems to be speculative so far.

One of those who refuse to connect the frequency of tropical storms with global warming is Roger Pielke Jr from the University of Colorado in Boulder. He points out that there has been a more intense hurricane activity in the Atlantic after 1995, but there are no reasons to think that it exceeds the variability seen in the past. Similar trends appeared in the mid twentieth century, with Florida hit eleven times between 1944 and 1950.

Still, the laws of physics discussed in the previous chapter make it clear that warmer oceans stimulate the formation of hurricanes. So if we cannot prove a link now, it is because the oceans have not reached high enough temperatures yet. Nevertheless, the number of intense hurricanes has increased in recent years, a fact expected to happen in a warmer climate.

So does the reliable evidence confirm climate change? After all there have been short periods of warming in the past, such as the one from about AD 900 to 1300, which allowed Eric the Red and his Vikings to settle in Greenland. Aren't we now experiencing a similar passing fluctuation?

Temperature measurement is crucial in answering this question, and scientists have checked the satellite record for the layer just above where airplanes fly. The results agreed with those obtained by weather balloons: they showed less temperature change in the tropics than the theory predicts. It's unclear whether this disagreement results from measurement flaws, incorrect models, or a combination of both, and scientists are struggling to solve the puzzle.

Or is this mismatch telling us that we have rushed into conclusions?

Complications with Chaos

In May 1989, the University of Massachusetts at Amherst hosted a conference dedicated to global warming. Its main objective was to assess the effect of greenhouse gases on climate change from observations and modelling. Among participants was Edward Lorenz, the scientist who had coined the term 'butterfly effect'.

I met Lorenz in 1998 after a public lecture he gave at the University of Victoria. The next day we lunched together at the Faculty Club. In his early eighties at that time, he was in good shape. His eyes were lively, and his interest in science seemed unabated. These impressions came back to me recently as I read his Amherst paper, which has the same standards, strength, and composure of his personality.

Lorenz used a model with twenty-seven variables, which included many physical processes, such as cloud formation and heat exchange. It was not an accurate reflection of the physical reality. He thought it didn't have to be. He was interested only in how atmospheric models behave in the absence of external influences. Had he used more complex simulations, he could not have run his routines long enough to cover several centuries of climate change. Computer capacity at the end of the 1980s was fairly modest.

As one of those who had rediscovered the chaos phenomenon, Lorenz was not surprised with his new findings. Sea-level temperatures showed oscillations of up to 2 degrees Celsius, with patterns of high or low regimes lasting from one to several decades. In spite of no external influences, the model led to spontaneous variations due to its chaotic character. So fluctuation between decades of higher and lower global temperatures might be just normal behaviour. 'Perhaps the best advice chaos theory can give us,' Lorenz remarked, 'is not to jump at conclusions.'

This exercise led him to an obvious question: If slight variations in temperature occur naturally, how can we tell when greenhouse gases affect the climate? The only way to provide an answer is through statistical analysis. Unfortunately, in 1989 there was little data. Consequently Lorenz left the question open. From observations alone, he could neither say that a greenhouse-induced warming had already set in, nor that it had not.

In the meantime, however, the IPCC (Intergovernmental Panel on Climate Change) experts succeeded in answering this question. Using their new models, which come much closer to reality than the one Lorenz used in 1989, they still detected chaos in the absence of external influences, but the temperature variations were very small. The average global temperature increase registered in the past few decades exceeds this variation several times.

Chaos, however, doesn't entirely leave the picture. The new models are still sensitive with respect to the initial conditions once external influences, such as carbon dioxide, are taken into consideration. Moreover, the stories various models tell are quite different from each other, and chaos is one of the culprits for this outcome. For instance, some models forecast an 8 degree Celsius increase in temperature by 2100 and others only 3 degrees, so the difference between them is higher than the minimum predicted.

Global-warming deniers claim therefore that this large margin of error renders these forecasts useless. Again, this statement is misleading. Though quantitative forecasts may not be exact, what matters is that no serious model predicts a decrease. All of them are consistent: greenhouse gases are warming the planet. Otherwise, we would see periods of warming followed by cooling even in the presence of carbon dioxide. Since no model predicts such behaviour, climatologists are very likely correct.

Chaos raises another problem. Tampering with the current conditions a little can lead us to large changes in the future.

In other words, a small quantity of greenhouse gases or aerosols emitted today has the potential to alter the climate to a state very different from the one it would otherwise reach. Therefore we don't know the outcome in the long run. To find out these things, mathematicians need to get involved. At the time I write these lines, the mathematical community has launched an appeal in this direction, so we are likely to have a better perspective on global warming in a few years from now. But the short-term future is quite clear unless something unexpected happens.

Indeed, the predictions discussed so far are likely to occur only if the current trend continues and no extreme events take place. Under special circumstances, the climate could change within years. Though improbable, this possibility cannot be overlooked. Research done in the early 1990s on three-kilometer-deep ice cores in Greenland show that abrupt changes happened several times in the past 100,000 years. So what could trigger such a catastrophe now?

Worst-case Scenarios

In December 2006, a team of scientists from three American universities presented some sensational conclusions at a press conference in San Francisco during the annual meeting of the American Geophysical Union. Even a regional war between two countries, each dropping a few Hiroshima-size bombs over the other, would have devastating consequences for the climate, they claimed. The 'nuclear winter' that would follow might last ten years, long enough to destroy agriculture, the world's economy, the financial institutions, and our social milieu, not to mention the millions of deaths and the effect on the entire biosphere.

In their calculations, these experts took into account the recorded changes due to large volcanic eruptions, which affected both the weather and the climate, and compared these effects with the ones atomic bombs would produce. So even if regarded solely from the point of view of climate change, humankind cannot afford a nuclear war.

Worse consequences would be felt if a large cosmic object hit Earth. The next chapter deals with the problem of detecting and preventing such a catastrophe. As in the nuclear-war scenario, climate change would be one of the few disastrous consequences of a cosmic impact which could wipe out most life on Earth.

A more immediate, but still unlikely, disaster might occur if the Gulf Stream were to slow down or stop. Scientists are quite moderate about this issue. The media, however, has made it become popular with the general public. Al Gore, the former American Vice-President, largely helped to spread this idea with his Oscar-winning documentary *An Inconvenient Truth,* which presents this ocean current in a vivid animation.

The idea that the Gulf Stream regulates Europe's mild climate originates from a book published in 1856 by an American army officer named Matthew Fontaine Maury. If a large volume of fresh water from melting ice mixed rapidly into the North Atlantic, the existing temperature balance would be upset and the Gulf Stream might slow down or stop. Deprived of the warm water coming from the equator, Europe would be plunged into permanent winter. Or so the popular belief holds.

The Day After Tomorrow, a 2004 Hollywood movie, takes this idea to the extreme with a story in which the transition from global warming to global cooling happens within hours. Though no scientist validates this scenario, some point out that the Gulf Stream has been disturbed more than once. They correlate one of these disruptions with an average temperature reduction of several degrees in Western Europe 12,000 years ago.

But not everybody links Europe's climate with the Gulf Stream. Richard Seager, a researcher with the Lamont–Doherty Earth Observatory at Columbia University, is one of those who play down this connection. His mathematical models show that the atmospheric transport exceeds by several times the influence of the Atlantic over Europe's climate. In his view, the air currents are responsible for the mild temperatures of Western Europe, not the ocean streams, whose influence is fairly negligible.

Seager concluded one of his articles with the following words: 'It is about time that the Gulf Stream-European climate myth was resigned to the graveyard of defunct misconceptions along with the Earth being flat and the Sun going around the Earth. In its place we need serious assessments of how changes in ocean circulation will impact [the] climate . . . and a new look at the problem of abrupt climate change that gives the tropical climate system and the atmosphere their due as the primary drivers of regional climates around the world.'

Seager's theory appears to make sense. The coastal regions of Western Canada, where I live, enjoy a mild climate due to a warm ocean current. This effect, however, is felt only near the shore. British Columbia's interior has the same harsh winters as the rest of the country. Were ocean currents as influential as the Gulf Stream is claimed to be, my entire province should have warmer winters than it has. But things may be not as simple as that. Climate physicists are still trying to solve this puzzle.

Although there is no agreement about the consequences of disrupting the Gulf Stream, most experts are worried that the rainforest of the Amazon is dying. Computer simulations done in the 1990s at the Hadley Centre for Climate Prediction and Research in Britain show that the collapse of this ecologic paradise would deprive us of a key oxygen producer. Consequently, the carbon dioxide in the atmosphere might go within decades as high as 1,000 ppm, two-and-a-half times the current level, leading to a significant increase of the average temperature.

Global warming affects many plant species, including those of the Amazon. Should the rainforest be in danger, as the experts have warned, we would have full evidence around 2040. But by then it will be too late to save this natural carbon-dioxide filter, which seems to play an important role in balancing Earth's climate.

In his book *The Weather Makers*, Tim Flannery describes another worst-case scenario: the release of methane from the sea floor. Huge volumes of this gas could emerge if the deep oceans get warm. Some scientists believe that 245 million years ago, 90 per cent of the pre-dinosaurian species became extinct because of this phenomenon. But though possible, the release of methane is unlikely. Several deadly asteroids could hit us before methane fills the air. Then what is the best bet for the Earth's future climate?

The Climate to Come

The past three decades have seen an intense activity for understanding climate mechanisms and for issuing predictions.

Progress has been good in both directions. But there are still many question marks, some of which this chapter has touched. Unfortunately we cannot experiment with the climate to check the results. Nevertheless, climatologists point at an incoming global warming, for which we seem to be responsible.

The most likely scenario, as outlined in the 2007 IPCC report, is that of average temperature increases of between 1.8 and 4 degrees Celsius by 2100 with a sea-level rise of about half a metre, assuming that we will gradually reduce greenhouse gas emissions. This reduction is not beyond our means. Economists calculate that, if we start acting now, the costs would be quite moderate compared to, say, the amount spent on gambling or advertising or what it takes to wage war in Iraq.

Not acting at all would impact the climate for centuries to come. To understand the consequences, let us compare the current situation with having already filled half a swimming pool with water. Even if we close the tap and pull the plug, it will still take some time until the pool grows empty.

On this point, my colleagues Andrew Weaver and Jeff Foss seem to agree. Though they differ about the reasons why the climate is warming, they both care about our environment. One doesn't need to be a scientist or a philosopher to understand that we cannot pollute nature forever, and we had better start addressing this problem now. The only real question we are facing is how fast we should pace our actions. The best strategy suggests starting with small changes right away and gradually increasing the pace in the future. Panicking would be in nobody's interest.

Indeed, many scientists are worried about the current fear global warming has created. Even if one of its positive consequences is that governments are starting to act, apocalyptic views are not serving us in the long run. At worst, they give science a bad name. Exaggerating climate change can be as damaging as denying it. Alas, some scientists fall into this trap, as I recently saw on television, when a researcher claimed that the oceans would definitely swell more than six metres at the end of the century.

This episode reminded me of a front-page article showing two maps of my city by 2100 in what an environmental group claimed to be the limits between which the oceans are going to rise: a minimum of six metres and a maximum of sixteen. The article mentioned Andrew Weaver, implying that my colleague had supported this view. The next day, the newspaper published a letter in which Weaver angrily denied those claims.

Weaver represents the majority of climatologists. Though convinced that our planet is warming, they remain honest in spite of the pressure various groups put on them. They give us the true results of their findings and suggest a reasonable solution to the problems we have. Even when the issue is highly politicized, they keep their ground against any partisan pressure. People like them make science credible.

Mike Hulme, the founding director of the Tyndall Centre for Climate Change in Britain and lead author of the chapter on climate-change scenarios for an IPCC report, had the following to say about these developments: 'Over the past few years, a new environmental phenomenon has been constructed in this country—the phenomenon of *catastrophic* climate change. The increasing use of this term ... has altered the public discourse, [which] is now characterized by phrases such as *irreversible tipping in the Earth*'s climate and *we are at the point of no return*.'

Hulme is convinced of global warming. Asked, however, to give a number for the maximum rise of the oceans, he said he didn't know. His colleague Carl Wunsch from the Massachusetts Institute of Technology offered more insight: 'Nobody can tell you what the probabilities are,' he said. 'The probability of another metre of sea-level rise in the next 50 years isn't zero, but it isn't 90 percent either. And if you pinned me down to tell you what it really is, I couldn't do that.' My colleague Andrew Weaver was less evasive: 'upper bound for the sea-level rise this century is a metre,' he said.

These answers characterize the nature of predictions climatologists can do these days. They provide a tendency and, under certain hypothesis, estimate future temperatures, ocean levels, and precipitation patterns. But none of them knows at what extent governments will protect the climate or whether a nuclear war will erupt.

Scientists, however, can offer plans to stop or reverse global warming. The IPCC has already mentioned some potential last-resort solutions: a 100-square-kilometre sun-blocking disk installed in space at a proper distance from Earth would reduce solar radiation; controlled scattering of sulphur in the air, as suggested by the Nobel Prize winner Paul Crutzen, would have a cooling effect; and the fertilization of oceans with iron, which stimulates the growth of the carbon-dioxide-hungry phytoplankton, would also help cool the planet.

Let's hope we will never need these measures, and the entire world will unite to fight this problem for the sake of future generations. The fate of the next century's climate is now in the hand of politicians. However, in democracies at least, *we* are the ones who elect them.

6. EARTH IN COLLISION: COSMIC IMPACTS

Old men and comets have been reverenced for the same reason: their long beards and pretences to foretell events.

Jonathan Swift

I have always found the night sky fascinating. As I child, I liked stargazing with my father from the balcony of our house. We often counted the meteors that flashed in the dark, and I was happy when we could beat our previous record.

In school I was interested in many subjects—history, languages, and literature among them—but I chose to study mathematics. I never regretted this decision. The pursuit of reasoning filled my life with composure, excitement, and joy, and helped me connect my work with my hobbies. Even the writing of this book is a natural extension of my day-to-day preoccupations.

This harmony was possible because of my main research domain, celestial mechanics, which lies where astronomy and mathematics overlap and is the kernel out of which the field of dynamical systems has grown. One of the problems I approached in my youth was that of collisions between cosmic bodies. My doctoral thesis explored some theoretical aspects of this topic. But the impact between a planet and a comet or an asteroid concerned me beyond mathematics. I had good reasons for that. Such events had happened in the past.

Jupiter in Collision

In July 1994, a spectacular event gripped the astronomical world. A trail of twenty-one cosmic objects, remnants of the comet Shoemaker–Levy, hit Jupiter at seven-hour intervals on average. During the six days this show lasted, the fragments slammed into the planet like a breaking string of pearls with speeds of 200,000 kilometres per hour. They released an amount of energy equivalent to more than two million Hiroshima bombs. The largest piece left a temporary scar in Jupiter's atmosphere as big as Earth.

Though unique in the history of astronomical observations, these collisions were predicted well in advance. On 25 March 1993, Carolyn Shoemaker was examining the images her husband Eugene, an astrogeologist, and David Levy, an amateur astronomer, had taken just hours earlier. Shot forty-five minutes apart from the Schmidt telescope at Mount Palomar in California, the photographs made clear that a dot had moved. The team had discovered a new comet.

On 26 March, an image taken by Jim Scotti from Spacewatch in Arizona showed the original dot consisted of five pieces. The same day, Brian Marsden of Harvard announced that the objects were probably not far from Jupiter. Better images taken later showed the comet was split into at least twenty-one nuclei. Mathematical calculations predicted a collision with Jupiter. So on 16 July 1994, when the first piece hit the planet, thousands of telescopes, including Hubble, witnessed the event.

This discovery made Levy and the Shoemakers famous. But Eugene couldn't enjoy his celebrity for long. In 1997 he died in a car accident in Australia while on a research trip to study impact craters. Carolyn survived the head-on impact their car suffered with another vehicle. Eugene Shoemaker was the first earthling whose ashes were scattered on the Moon, a mission accomplished by a lunar spacecraft two years later.

Widely covered by the media and watched all over the world, the collision between Jupiter and the comet Shoemaker–Levy raised public awareness about the consequences of an impact with Earth. Since the Hiroshima bomb killed 140,000 people, it is not difficult to imagine what the equivalent of two million blasts of that kind would produce. Yet, too little is done these days for predicting and preventing such a catastrophe.

Rocks From the Sky

Towards the end of the eighteenth century, a physicist from Saxony named Ernest Chladni linked sound waves with the patterns of sand strewn on a vibrating plate. For this idea, he is remembered as the father of acoustics. But some of his peers thought lowly of him. Georg Lichtenberg, a Göttingen professor known for his work in electricity, described Chladni as 'hit on the head with one of his stones'.

Lichtenberg alluded to a slim volume entitled *On the Origin of Iron Masses*, which Chladni had published in Riga in 1794. In this book, Chladni claimed that stones that fall from the sky cause the fireballs seen in the atmosphere. He believed in the cosmic origin of those rocks, which probably came from the debris of a celestial collision or explosion.

As Lichtenberg's quote shows, Chladni's book was met not only with scepticism but also with scorn. Scientists opposed the idea that rocks could fall from the heavens. Their reluctance persisted even when they were presented with pieces of meteorites and faced with independent witness reports.

Only two months after the book was published, meteorites fell with noise before the eyes of several villagers from Cosona, close to Siena, in Italy. Lucrezia Scartelli, a woman who witnessed the event, picked up a pebble-sized stone but dropped it quickly because it was too hot. Ferdinand Sguazzini, another eyewitness, saw bigger rocks going deep into the ground. At the same time, people living kilometres away heard artillery-like noises coming from a clear sky.

How could this phenomenon happen? The explanation that won acceptance belonged to Lazzaro Spallanzani, a leading Italian volcanologist briefly mentioned in Chapter 3. He claimed that a strong wind had carried those rocks from Vesuvius, which had erupted eighteen hours earlier some 320 kilometres southeast from Cosona.

Ambrogio Sodani, a geology professor at the University of Siena, was one of the few who disagreed. He researched the region and concluded that those peculiar stones, which had spread over an area of about fifty square kilometres, contained mostly iron and were very unlike the rocks ejected from Vesuvius. In Sodani's view, the stones had cosmic origin, but his opinion was largely ignored. The scientific world was not ready to accept this conclusion.

It took a few more such events, witness reports, and collections of samples, until Chladni's explanation was accepted. Stones fell from the heavens at Wold Cottage, England, in 1795; Evora, Portugal, in 1796; Benares, India, in 1798; and L'Aigle, France, in 1803. No volcanoes erupted in the affected areas before those occasions and no strong winds blew. The admission that meteorites had cosmic origin came in the report of the physicist and mathematician Jean Biot, who had studied the incident at L'Aigle for the French Academy.

When Biot reached his conclusion, astronomers were living through exciting times. Without knowing it, they had discovered what was responsible for the falling rocks.

Minor Planets

In 1772 Johann Elert Bode, the German astronomer who suggested the name of Uranus, published an empirical formula that gave the approximate distance from the Sun to each planet of the solar system. His countryman Johann Daniel Titius had already obtained that result six years earlier, but Bode didn't credit him. Nevertheless, their contribution is known today as the Titius–Bode law.

The formula, however, failed in one case. It asked that a planet revolve around the Sun between the orbits of Mars and Jupiter, where nobody had ever seen a cosmic object. The Titius–Bode law was an incentive to find the missing wanderer, which many telescopes began to seek.

For almost three decades the search bore no fruit. But on 1 January 1801, the Italian astronomer Giuseppe Piazzi discovered a new cosmic object, which was faint and of Jupiter's colour. Piazzi suspected he had found a yet unknown star. After three more nights of observation, however, he changed his mind. The object was moving. Piazzi continued to follow it until an illness forced him to quit on 11 February. As he resumed his observations, the object was nowhere to be found.

Piazzi was cautious when he revealed his discovery. 'I have announced this star as a comet,' he wrote to his colleague Barnaba Oriani of Milan, 'but since it is not escorted by any haze and its movement is slow and fairly uniform, it has occurred to me several times that it might be something better than a comet. But I have been careful not to advance this supposition to the public.'

To prove he had discovered the much-sought planet, Piazzi had to determine its orbit. Unfortunately he didn't know how to compute it from the few observations he had made. There were no known mathematical methods that could help him solve this problem. Without an orbit, however, he couldn't recover the object. Piazzi saw no exit from this loop.

Several astronomers, including Bode, believed Piazzi had found the missing planet. They spread the news in the hope that someone would retrieve it. During the following months, several experts tried to compute the trajectory from Piazzi's observations, but all of them failed. No astronomer saw the new cosmic object in the regions of the sky suggested by those calculations.

Luckily, a young German from Göttingen named Carl Friedrich Gauss, who would become one of the world's greatest mathematicians of all times, learned about the problem and solved it by inventing the method of least squares. When, on

7 December 1801, the telescopes pointed in the direction Gauss had indicated, the lost object was there. Further observations confirmed it to be a small planet moving between the orbits of Mars and Jupiter. They called it Ceres, after the Roman goddess of fertility.

Many thought that Ceres was the missing planet predicted by the Titius–Bode law, but some astronomers had doubts because the object was too small to show a planetary disk. The group of sceptics grew after 28 March 1802 when another object, Pallas, was discovered. Gauss showed Pallas to be at the same distance from the Sun as Ceres. Similar objects were discovered in 1804, Juno, and 1807, Vesta. These findings proved the sceptics right. The new celestial bodies were named asteroids or minor planets.

The search for them didn't stop here. The thousandth— named Piazzia, to honour Ceres's discoverer—was found in 1923. Current estimates put the number of asteroids larger than one kilometre to between 1.1 and 1.9 million. There are many more small asteroids, and they are responsible for the stones that fall from the sky. The terrestrial impact of a seventy-five-metre-wide object would have an effect similar to the Tunguska explosion, which equated 1,000 Hiroshima bombs.

Much has been written about this mysterious event that took place in the Central Siberian Plateau near the Stony Tunguska River on 30 June 1908. A fireball appeared in the early morning sky, moving within a few seconds from south-southeast to north-northwest and leaving a long trail of light. The object fell for a few minutes and exploded eight kilometres above the ground, raising a huge mushroom of dust. More than 2,000 square kilometres of forest were flattened.

Various theories have been proposed for this phenomenon, from a matter–antimatter crash to the passage of a black hole through Earth. The most plausible explanation is Earth's collision with a comet or, more likely, an asteroid not larger than a baseball field. But while an object of this size does not threaten our existence as a species, a ten-kilometre-wide stone could kill people in the hundreds of millions and cause major extinctions on Earth.

The Death of the Dinosaurs

In 1977, Walter Alvarez, a geologist at the University of California at Berkeley, discovered that some clay samples he had dug out in a survey near Florence in Italy contained thirty times more iridium than normal. Iridium is rare in Earth's crust but common in asteroids. The samples came from a geological layer at the border between the ages of the dinosaurs and the mammals.

After discussing his findings with his father, Luis, a Nobel Prize laureate in physics, Alvarez came up with a crazy idea. What if the iridium he found had resulted from the impact of an asteroid with Earth sixty-five million years ago? His hypothesis could explain the extinction of two thirds of the species living at that time, including the dinosaurs, and the subsequent spread of the mammals.

Most of the living beings that disappeared didn't have to die at impact. Collateral events, like forest fires and dust spread in the atmosphere, disrupted the climate and the ecological balance. Without enough light for photosynthesis, most plants died. The majority of animals did not survive long without sufficient food.

To prove his theory, Alvarez had to find the impact crater. He and his father computed from the quantity of iridium they had sampled that the asteroid should have been nine kilometres in diameter or larger. Such an object would have produced a crater at least 180 kilometres wide, not necessarily close to the excavation site. Due to the immense explosion, the iridium would have dispersed in the atmosphere before settling down in the geological layer where Alvarez found it. But was there such a crater?

Most scientists viewed this theory with scepticism. Some, however, found the logic correct, and provided skirting evidence, such as the survival of some aquatic algae in spite of the extinction of most marine life. Still, there was a weak link. How could some dinosaurs survive for years with no plants to eat? The defenders of the theory answered that not all the plants died at once. The extinction process must have been slow and painful. But without a large enough crater formed at the right time, they lacked proof.

The breakthrough came in the 1980s, when a multi-ring crater was discovered in Mexico's Yucatan peninsula. Its existence could be proved only through geological surveys made under a kilometre of post-impact sediment. The age matched too: the crater was sixty-five million years old. But there was a little problem. The third (and believed to be the largest) ring should have been a bit larger to match the theory.

There is no unanimous agreement about how multi-rings originate. An appealing hypothesis compares their formation to the waves a stone makes when thrown into a lake. When a high-speed object hits Earth, the rings 'freeze' instead of vanishing. In the framework of this theory, the largest ring of an impact crater of such depth must be more than twenty times the size of the object that creates it. Measurements matched only barely in Mexico. To account for all the iridium found in samples, the asteroid had to be nine kilometres in diameter but preferably larger, whereas the third ring was only 180 kilometres wide.

The final piece of the puzzle was found in 1993 when a study of gravitational anomalies in the region of the crater led to the discovery of a fourth ring, which is almost 300 kilometres wide. More recent studies, however, dismiss the necessity of the multiring theory for proving the extinction of the dinosaurs. They claim that the volume of post-impact debris accounts for the megadisaster. Independently of such details, scientists are now confident that life on Earth has been affected by an impact with a minor planet. And astronomers know that we will be on a collision course again.

Near Hits

On 23 March 1989 an asteroid known as 1989 FC passed within 700,000 kilometres from Earth, almost twice as far as the Moon. Those who consider this distance large enough not to worry should think that the object missed us by a few hours of space travel in terms of its speed relative to Earth.

Asteroid 1989 FC was 500 metres wide and passed us by at more than 60,000 kilometres per hour. It was too large to explode in the atmosphere, so had it hit Earth on land, it would have dug a ten-kilometre-wide crater. Had it fallen into the ocean, it would have caused a tsunami about 100 meters tall. In either case, the consequences are beyond imagination. But more disturbing than the thought of the disaster is that the culprit was found three weeks after it passed Earth. A hit would have taken us by total surprise.

Asteroid 1989 FC was not the most dangerous in recent history, but it swayed the American Congress to target small amounts of money for tracking rogue cosmic objects and preventing an impact with them. An even closer call occurred on 19 May 1996, when the minor planet 1996 JA1 passed us by at a distance of 450,000 km, or just a bit farther than the Moon. This asteroid had about the same size as 1989 FC.

Dangerous approaches have happened earlier, but they have had no effect on government decisions. A spectacular near hit occurred on 26 June 1949, when a photographic plate taken with a telescope at Mount Palomar in California recorded the long trail of a fast asteroid. The image registered the discovery of Icarus, a minor planet 1.4 kilometres in diameter, whose eccentric path crosses the orbits of Earth, Mars, Venus, and Mercury.

Icarus's orbit could be determined with no difficulty. This object needs a bit over a year to go once around the Sun. In 1968 it came within six million kilometres from us. The next close approach will be in June 2015, but there is no concern for a collision in the foreseeable future.

In 1967, after the near hit of Icarus had been predicted, professor Paul Sandroff from the Massachusetts Institute of Technology asked his students to find a way to destroy this minor planet in case it came on a collision course with us. Known as *Project Icarus*, the story of how they conceived their plan was dramatized in the 1979 science-fiction movie *Meteor*, starring Sean Connery.

These examples show how important it is to monitor minor planets like Icarus, even if they do not pose a danger now. They could collide with smaller objects, whose orbits we don't know, break into pieces, and threaten our existence. Indeed, some recent computations point out that the asteroid that killed the dinosaurs resulted from the collision between two large asteroids, one 170 and the other sixty kilometres wide.

Small objects, a few metres wide at most, hit Earth quite often. The media reports from time to time how people see and hear blow-ups in the sky. If the events take place at night, the descriptions are as colourful as the images they witness. Such explosions produce rocks, which could damage a roof or two, but most often have no serious consequences. The chance of being hit by such a stone is much smaller than that of being struck by lightning. Large objects, however, pose a serious threat to life on Earth.

Comet Alert

In 1877 the French writer Jules Verne published *Off On A Comet*, which is one of his strangest and most polemical books. Unlike most of his novels, this one ignores the achievements of science and gives free rein to his imagination. The story tells how a comet grazes Earth, pulling away a piece of North Africa along with the novel's main characters. After it circles the Sun, the comet comes back to Earth to return the heroes and the piece of land.

A real encounter with a comet would be nowhere near that romantic. Depending on its size and the spot it hits, the consequences of an impact could be devastating. And we have evidence that such collisions with Earth took place in the past more often than we would like to believe.

In 1984 paleontologists David Raup and John Sepkoski from the University of Chicago published a study on the fossil record found on Earth. The conclusion of their analysis is astounding: major extinctions occur about every twenty-six million years. Several theories tried to explain this phenomenon. One of them, proposed by Marc Davis and Richard Müller from the University of California at Berkeley and Piet Hut of the Institute for Advanced Study in Princeton, claimed that this almost periodic extinction takes place because comets hit us every time.

Comets are thought to come from the Oort cloud, which is a postulated spherical region around the Sun with a one-light-year radius. Davis, Müller, and Hut suggested the existence of a companion star, called Nemesis, which orbits the Sun once every twenty-six million years. When it reaches the Oort cloud, Nemesis pushes millions of comets in the solar system, at least one of which collides with Earth.

But why did these experts only suggest the existence of Nemesis without pointing at it, for a star so close to us must be visible? They answered that Nemesis must be among the thousands of catalogued stars whose distance was not yet determined. After six years of research, however, Müller found no trace of Nemesis and admitted that it may not exist.

Another theory claims that the disk of our solar system wobbles in the galaxy with a period of about sixty-four million years. So at half time the solar system and the galaxy are in the same plane. Then the galaxy's pull on the Oort cloud sends a comet shower towards the solar system. Since the period is approximate, it could perhaps explain the twenty-six-million-year cycle suggested by the fossil record.

Richard Müller of the Nemesis hypothesis and Walter Alvarez, who had explained the death of the dinosaurs, advanced another theory. They examined records of large impact craters on Earth and reached the conclusion that most of them occur in a 28.4-million-year cycle. Within measurement errors, this period is likely to agree with the fossil records. The probability that such collisions are accidental is very small.

There are other theories that try to explain the physical evidence of collisions with comets at time intervals that are not necessarily periodic but not random either. Certain, however, is that such events happen. Of course, the thought of a cosmic impact millions of years from now would not make us lose sleep, but a comet could hit us anytime with a warning of a couple of years to a few months. Unless we stand up to the challenge, we could have the fate of the dinosaurs.

What Can Be Done?

Explosions of Tunguska's magnitude could happen as often as one per century. They do not threaten to destroy life on Earth, but they could kill millions of people. Things would be different if a larger object hits us at high speed. Luckily, such events are rare, and our very existence proves that no major collision has occurred since the human species has evolved. The laws of probability, however, teach us that such events will take place in the future. So the problem is not *if* but *when*.

Until recently, our civilization was not advanced enough to prevent an impact with a large asteroid or comet. Today science and technology provide us with means to defend ourselves. So what can we do to stop an eventual cataclysm that could lead us close to extinction?

First we have to watch the sky and keep track of the objects that could hit us. This process involves observations and computations. None, however, is without difficulties, as we will further explain. Then we must find a way to dodge the impact. Several solutions were proposed, some easier to achieve than others. All of them, however, are costly and need to be verified in practice. But before getting into prevention techniques, let us see how collisions are predicted.

Space Watch

Composed mostly of ice and dust particles, comets were born in the outskirts of the solar system, which is about 4.6 billion years old. The giant outer planets, Jupiter, Saturn, Uranus, and Neptune, came into existence from agglomerations of billions of comets. Asteroids formed closer to us between the orbits of Mars and Jupiter, and are leftovers from the initial accumulation of the inner planets Mercury, Venus, Earth, and Mars.

The comets and asteroids that could eventually hit us are called near-Earth objects, or NEOs, because their orbits come close to our planet. There are several hundred known NEOs one kilometre in size or larger, and the list is growing fast. Their total number is estimated to be between 1,000 and 1,200. So far, astronomers have catalogued about 90 per cent of them, as well as more than 4,000 NEOs of all sizes.

The vast majority of NEOs are minor planets, referred to as near-Earth asteroids, or NEAs. Astronomers divide them into groups, depending on the shape of their orbits. They are the Atens, Apollos, and Amors, named after the best-known asteroid in each group. The near-Earth comets, or NECs, are classified into long- and short-term period, depending on whether they need more or less than 200 years to return to the Sun.

In the late 1980s, the science establishment and the media pressured various governments to do something about the rogue objects that could hit us. The publicizing of the 1989 FC occurrence had raised awareness in the minds of the general public. In the early 1990s, the American Congress approved funds for research aimed at finding all NEOs and proposing solutions to prevent potential collisions.

There are currently seven observatories that seek NEOs: five in the United States (in New Mexico, Hawaii, and three in Arizona) as well as in Italy and Japan. All the NEOs are published on a daily updated website. If the pioneering work started with photographic plates, the technically superior *charged coupled devices*, or CCDs, make detection easier since 1985. The CCD cameras take digital images like the camcorders. Computers help then with the calculations.

While the CCD technology is more accurate than the photographic method, the discovery technique hasn't changed. The experts take three or more CCD pictures of the same sky region at several-minute intervals. Then they compare the images to see if any NEOs have moved. The brightness, the distance travelled, and the direction of motion are helpful in identifying how close the object is to us. A dot that moves rapidly from one image to the next represents almost certainly an intruder.

These seven observatories are not the only ones finding NEOs. Various research institutes that study asteroids and comets also discover new objects, some of which come close to Earth. One of the foremost experts in the field is Brian Marsden, an astronomer based at Harvard University, where he directs the Minor Planets Center.

In 1998 I had the opportunity to meet Brian at a conference in celestial mechanics held in Namur, Belgium. He was cordial and energetic, and I very much enjoyed his company. Some of the details on how asteroids are discovered and tracked, I learned from him. This is when I found out that the space-watch initiative is not perfect. The observatories located on Earth could fail to see some NEOs, namely those that come towards us from the Sun's direction. Asteroid 1989 FC was exactly of this type, and we detected it three weeks after it passed us by. Had it been smaller, we might have missed it completely.

To fill this gap, several experts have proposed the launching of space telescopes, which could scrutinize the sky in the direction of the Sun. On Earth the atmosphere diffuses the light, but from a few hundred kilometres away the sky looks black even in the Sun's direction, so NEOs hiding there would be easy to detect. The main difficulty with launching such a project is financial, but given the low cost of the current space-watch programme and the tremendous loss we'd suffer if we failed to predict a future impact, this effort would be justified.

Computing Trajectories

In 1687 Isaac Newton published his masterpiece. It was entitled *Mathematical Principles of Natural Philosophy*, a work better known as *Principia*. In this book, he laid the foundations of celestial mechanics, a field of science that aims to predict how cosmic objects move in space. Newton's initial goal was to understand the motion of the Moon, but his ideas could be extended to all celestial bodies, including planets and stars.

Newton's work allowed the formulation of the N-body problem, which seeks to determine the past and future positions of N celestial bodies among which the gravitational force acts. The case of two bodies ($N = 2$), also called the 'Kepler problem', is relatively easy to solve, and Newton dealt with it in *Principia*. The orbit of one body relative to the other could be a circle, an ellipse, a parabola, a hyperbola, or a straight line (Figure 6.1). For three or more bodies, however, the problem has not been solved in general.

Nevertheless, the practical aspects of the solar system can be approached with so-called numerical methods, which offer approximate solutions to the equations that describe celestial motions. In fact, the eight planets, Mercury, Venus, Earth, Mars, Jupiter, Saturn, Uranus, and Neptune, together with the Sun form a nine-body problem, whose numerical solution is reliable for long intervals of time. We can predict the positions of these objects millions of years from now, and are confident that no collisions will occur between any of them.

Things, however, are more complicated with asteroids and comets. For several reasons, their orbits are difficult to compute

(a)

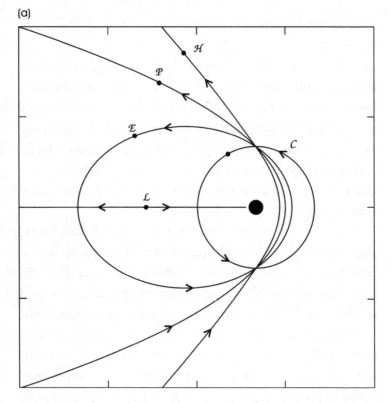

Figure 6.1 In the two-body problem, the orbit of one body relative to the other can be a circle, *C*; an ellipse, *E*; a parabola, *P*; a hyperbola, *H*; or a straight line, *L*.

with good accuracy soon after they are discovered. For instance, the masses are not known and the optical determination of the orbital elements is uncertain. When they discover a new such object, the experts mostly guess what value to attach to the mass and the initial conditions. But they correct their assumptions after many observations. Precise orbits can be obtained within years, yet comets are more elusive because their masses may vary. Therefore computed near-hits could be real collisions or vice versa.

A case that generated a lot of excitement was that of the two-kilometre-wide asteroid 1997 XF11, discovered in December 1997 by Jim Scotti of the University of Arizona. The observations of two Japanese amateur astronomers during the following two weeks allowed a preliminary estimate of the orbit, which proved to come close to Earth, with a dangerous approach in October 2028. The media picked up the news and claimed that a collision would be likely in three decades.

To improve their result, astronomers checked the photographic archives and found out that 1997 XF11 had appeared in some images taken in 1990, when nobody had recognized the dot on the plates to be an asteroid. But with a preliminary orbit at hand, the identification was easy. These images allowed the computation of a more precise orbit, which showed that, depending on how close 1997 XF11 will come to Earth in 2028, its passages in 2033, 2035, and 2045 could be as near as 15,000 kilometres, which is less than half the distance to the geostationary satellites. This minor planet was therefore added to the list of objects to be monitored from now on.

Future observations are necessary because 1997 XF11 might change its path if it gets close to some yet undiscovered comet or asteroid. The same remark applies to all NEOs. Regarding the problem from this angle, we are not safe at all. Therefore we must get ready to plan how to prevent a collision once we know that it might happen.

But if this asteroid doesn't pose an immediate threat, there is at least another object that does: the 500-metre asteroid Apophis, which will make a close flyby in 2029 and might collide with us in 2036. The level of alert was raised and lowered several times since Apophis was discovered in June 2004, and it looks like we will be able to estimate the impact chances better only after the 2029 passage.

Another reason why we must keep watching the sky is that we can see only 100–200 years into the future behaviour of NEOs. Due to the chaos phenomenon, which acts in the solar system, the orbits of these objects can change in time. But once dangerous objects have been tracked, keeping an eye on them should not be difficult for the next generations. It may well happen that some objects could be removed and others added to the list. This update, however, will require minimum effort.

Space Guard

Predicting a collision with an asteroid or a comet is a necessary first step to protect our planet. But it is not enough. If we don't find a way to prevent the disaster, all we achieve is to learn when the current cycle of life on Earth might end. Luckily, our civilization is advanced enough to prevent such a catastrophe.

The first proposals for avoiding terrestrial collisions came from science-fiction writers. Their imagination was often richer than their understanding of the scientific and technological level of the time. Nevertheless, the solutions they imagined were not completely out of touch with reality, and scientists could build on those ideas. So far, however, none of the proposed solutions has been checked, so they are now at an early stage of development.

A solution that might come to everyone's mind is to use nuclear weapons. The arms race that followed the end of World War II has seen a worldwide accumulation of nuclear heads, which could now be used to save us. The proponents of this plan think that if an impact is predicted one or two years in advance, we should launch a mission to crush the rogue object to rubble.

In 1992 Thomas Ahrens and Alan Harris of the California Institute of Technology reviewed this idea for various kinds of NEOs, from those no bigger than a football stadium to objects ten kilometres wide. They proposed a nuclear explosion involving a detonation near the asteroid to push the object along a trajectory away from Earth.

But the example of comet Shoemaker–Levy leads to second thoughts about trying to use nuclear means. We run the risk that the explosion splits the asteroid into a few pieces instead of crushing it to rubble, a situation that would lead to a scenario similar to the one we saw on Jupiter. Instead of suffering one blow, we could experience a bombardment that would hit several spots due to Earth's rotation around its axis.

Even if only one of those pieces or, better still, some debris reach us, their nuclear contamination could have devastating consequences for any form of life. Nuclear material is not something to tamper with. Imagine the consequences of a rocket exploding in the atmosphere a minute or so after launch, as happened to the space shuttle Challenger in 1986.

Experts in celestial mechanics are aware of another problem: dealing with more rather than fewer bodies. The orbits of the pieces formed after the explosion could get out of hand. To compute them might take longer than the time to impact. So unless the original object is small, and its pieces pose no real threat to life on Earth, the blow-up approach may not be the best idea.

A more ingenious way to crush the object would be to put in its path a net, similar to a builder's scaffolding, formed by millions of tungsten bulbs. The impact would create enough heat to smash the object to rubble. A spaceship could both install the framework in space and supply it with the energy needed to stop the intruder.

A safer and cheaper solution is to push the rogue object away from its collision course with Earth. If the impact is predicted two or three decades in advance, the steering can be done without much effort. For instance, we could send a space mission to hit the asteroid sideways and make it deviate from its trajectory by a few centimetres per second. Thus at the predicted collision time, the minor planet would pass several thousand kilometres away from Earth.

Of course, this is a simplistic example because the orbit of the asteroid is not a straight line, and in reality the computations become more complicated. But the idea is the same: by acting early with minimum effort, we force the object to pass at a safe distance from Earth. Experts in celestial mechanics can provide a good estimate of the direction and the amount of deviation that must be applied to get the desired effect, and engineers can plan how to achieve that goal.

In fact, the best way to change the path of the object is to push it from behind. A small increase in speed yields the same result as sideways nudging. The difference is that the rocket sent for this purpose can be smaller, cheaper, but as effective as the one aimed at the other mission.

Scientists have also imagined an alternative pushing mechanism. Instead of hitting the object once, they thought of implanting within it one or several rockets that expel jets of plasma. Such a device would work like a tiny rocket engine, which can be activated anytime to correct the path of the rock. Thus the asteroid or the comet would act like an unmanned spacecraft steered through sequential nudges.

A different solution is based on the fact that asteroids radiate heat after the Sun warms their surface. The heat emission gives the object a slight momentum, which translates into a push. So the object's molecular activity affects its course. A thin layer of black dust spread over the asteroid's surface would increase its temperature. White dust would lower it. Alternatively, a solar sail spacecraft could 'wrap' the asteroid in aluminium foil to affect the heat transfer.

This idea could be taken to extreme and used to generate a jet of material moving off the asteroid's surface. The effect could be achieved by deploying a huge parabolic aluminium mirror close to the asteroid. A ray of light focused on the object would vaporize a small part of it, affect the momentum, and lead to a change of trajectory.

The problem with this solution is that the mirror must be very large: 800 metres in diameter to deflect a three-kilometre-wide asteroid. Even though we are unable to place such a structure in space today, the times when we could do it are hopefully not far. Such a mirror could be built of pieces. The experience with the Keck telescope in Hawaii, which consists of thirty-six hexagonal parts, is a start in this direction.

These are some of the more realistic ways to prevent a future collision with an NEO. But even the ideas that belong to science fiction today might become achievable tomorrow. What matters most is the self-assurance that we can defend ourselves from the whims of cosmic motion. And we can, indeed.

Tom Gehrels, an astronomer who discovered more than 3,000 asteroids and several comets, and helped set up the space-watch programme for finding NEOs, outlined our current situation in the following words: 'Comets and asteroids remind me of Shiva, the Hindu deity who destroys and recreates. These celestial bodies allowed life to be born, but they also killed our predecessors, the dinosaurs. Now, for the first time, Earth's inhabitants have acquired the ability to envision their own extinction—and the power to stop this cycle of destruction and creation.'

Hopefully we will be spared a serious cosmic impact before we find all the rogue objects that wander in the sky and learn how to defend Earth against them. We should get there in a few decades, or sooner—if more resources are assigned to space research. Meanwhile, let's keep our fingers crossed.

7. ECONOMIC BREAKDOWN: FINANCIAL CRASHES

The only function of economic forecasting is to make astrology look respectable.

John Kenneth Galbraith

I was born and raised in Romania during its dictatorship, which lasted more than four decades. The December 1989 revolution marked the end of the wicked regime and the death of Nicolae Ceauşescu. Since then, the country has tried to adapt to the free market—a painful process for many of its citizens. But in time they have adjusted to change, and Romania joined the European Union on 1 January 2007.

I didn't witness those developments because I had defected to the West well before the revolution. Western democracy proved close to my nature, and I adapted quickly to its values. Still, it took me a while to understand how it works. The idea of investing in the stock market, for instance, was new to me. Private business had been illegal in Romania. If you saved money, you kept it in the bank.

A colleague in Victoria showed me the ropes of investment. From him I learned about the NASDAQ and the Dow Jones, and he lent me a book about the stock exchange. He had made millions of dollars through trading, so I trusted his advice. When he told me that a certain stock in British Columbia was hot, I bought a few shares right away. He was so convinced about their value that he put all his profits into them.

The stock's price went up indeed. It doubled in a week and kept hiking. I checked it on the Internet several times a day, amazed at how easy it was to make money. A month later my investment had tripled. 'What an amazing mechanism!' I thought. But my enthusiasm faded soon. One morning when I checked the stock again, its value had dropped to about half of what I had paid for it. From then on it kept falling. The company went bankrupt before the year's end.

My loss had been small, but my colleague was almost ruined. Worse: the relatives and friends he had advised to buy those shares were blaming him now. Luckily his pension fund was untouched, and he found solace in that thought.

This episode taught me a good lesson. From then on I wanted to learn more about how the financial world functions and what triggers the rise and fall of stock prices. Soon I realized that nobody is safe from a stock-market crash, whose consequences may affect investors and non-investors alike. A look into the past proved convincing.

Black Weekdays

In the early days of September 1929, the Dow Jones Industrial Average, which measures the overall condition of the New York Stock Exchange, reached a high of 381.17 points. This was great news for investors. The bull market they were experiencing since the beginning of the decade showed no signs of slowdown. Soon, however, their enthusiasm received a hard blow.

On 24 October—a day known as Black Thursday—stock exchanges in North America seemed to collapse. Panic set in. The investors wanted to sell their shares without even knowing what they were worth. A daily trade-volume of four million shares was deemed high then, but a 12.9-million record was reached that day.

By 12.30 p.m. the Chicago and Buffalo Exchanges closed down. Several known speculators had already committed suicide. In New York, some leading Wall Street bankers met at 1.00 p.m. to find a way to stop the panic and contain the damage. They chose Richard Whitney, the vice-president of the Exchange, to act on their behalf. There was no time to waste.

At 1.30 p.m., when Whitney showed up on the floor, the crowd went silent. They expected him to close the Exchange for the day, but were surprised to hear him ask the current price of US Steel, whose shares had dropped together with almost everything else. When a voice answered '195', Whitney offered to buy 10,000 shares at 205. Then he made similar offers for more than a dozen other stocks.

Miraculously, panic was contained. Investors began to buy, hoping to profit from this new surge. The market recovered a bit, but closed in the negative due to the massive morning sellout. Still, the prospects looked good. On Friday and Saturday prices proved steady. This cautious optimism, however, didn't last long. The weekend press covered the events in detail, and many investors decided to sell their shares.

On 28 October, known as Black Monday, the Dow Jones lost a record 13 per cent. Amid rumours that President Hoover would not veto a tariff act that increased taxes on foreign goods, prices kept falling on Tuesday. Rockefeller and other tycoons bought stocks to show their trust in the market, but their gesture didn't help. More than 16.4 million shares were traded that day, with a market drop comparable to Monday's. By mid November, the Dow had lost 40 per cent from its September peak.

The market recovered during the early part of 1930, but later that year another crash took place. From then on prices declined steadily until they reached a low for the century on 8 July 1932, when the Dow Jones registered 41.22 points. Counting inflation, the September 1929 peak was hit again only in 1954, a quarter of a century later.

Black Thursday marked the beginning of the Great Depression. The entire world suffered, with levels of homelessness and unemployment never encountered before. Industrialized countries were affected most. Rural areas also got hurt when crop prices fell up to 60 per cent. In a context of geopolitical instability, this crisis helped Hitler become Chancellor in Germany, an event that led to World War II and more than sixty million deaths. No conflagration had ever claimed so many lives.

It would be wrong to blame the collapse of Wall Street for all this evil. Nevertheless, the experts accept that the 1929 crash and the market decline that followed are among the factors that led to the Great Depression. Had this crisis happened at a time of economic strength, its impact might have been smaller. But given the circumstances, which included a bad corporate and banking structure, a dubious state of the foreign balance, and an uneven distribution of income, the crash mattered. So it is normal to ask whether such downfalls are predictable and what can be done to prevent them.

The Rules of the Game

One might be tempted to say that the growth and fall of the market is purely psychological, and people buy or sell shares depending on how they feel. For instance, the colleague who showed me the ropes of investment decided to put all his eggs in one basket, in spite of knowing the risks. Or, panic set in on Black Thursday until a respected banker broke the spell. Examples like these abound, so psychology cannot be neglected. But is this factor the dominant one?

Mathematicians think it is only part of the answer. They look at the stock trade in the framework of game theory—a branch of mathematics and economics that analyses how players maximize their gains. Though game problems are as old as history and can be found in ancient texts like the Talmud, a systematic development of the field started only in the twentieth century. The founder was John von Neumann, who was briefly mentioned in the chapter on tsunamis.

In 1928 von Neumann published the article *Zur Theorie der Gesellschaftsspiele* (On the Theory of Social Games), followed in 1944 by a book with Oskar Morgenstern: *Theory of Games and Economic Behavior*. Several other mathematicians, including John Nash of the book and movie *A Beautiful Mind*, followed with other contributions. It soon became clear that this theory helps clarify how people make decisions.

In this framework, the stock market is a particular game. Players aim to profit through trading shares in companies listed at the stock exchange. Every investor's dream is to buy stocks at their lowest price and sell them at the highest. But nobody can recognize the instants when the extremes are reached.

In the 1960s, a mathematician and electrical engineer named Claude Shannon, who had already become famous for laying the foundations of information theory, started to use game-theoretical principles in blackjack and roulette. He and his wife, Betty, went on weekends to Las Vegas to check the validity of the theories he had developed with physicist John Larry Kelly Jr, a Bell Labs associate. In no time, Shannon made a lot of money by playing honestly but with a certain strategy in mind.

He had based his approach on what is known today as the Kelly criterion, a formula that maximizes the long-term growth of repeated gambles. For instance, with the help of the formula one can compute that in a game with a 40 per cent chance of winning in which the player has 2:1 winning odds, a bet of 10 per cent of the gambler's bankroll will offer the best cumulated profit after sufficiently many bets.

Later Shannon applied similar principles to the stock market with even better results. His long-term portfolio gave returns of 28 per cent a year, better than most professional investment firms of his time. Indeed, in 1986 *Barron's*, a weekly Wall Street magazine, published an article ranking the performance of America's top seventy-seven money managers. Had he been mentioned there, Shannon would have ranked fourth. But what was the secret of his success?

Shannon did not think he had a secret formula that met the rigour of his mathematics. For this reason, he never published any paper on this subject. But he agreed to cooperate when others wanted to write about his investment strategies. A recent book, *Fortune's* Formula by William Poundstone, sheds some light into how Shannon maximized his returns.

As the founder of information theory, Shannon had a feeling for what data to pick or ignore. 'I think that the technicians who work so much with price charts,' he said, 'are working with what I would call a very noisy reproduction of the important data.' By noise he meant unnecessary information, which obstructs the assessment of a stock. 'The key data is, in my view, not how much the stock price has changed in the last few days or months, but how the earnings have changed in the past few years.'

Shannon plotted earning graphs for various companies and tried to estimate future returns based on past performance and current information. When possible, he checked out the viability of each company's products. For instance, when thinking whether to buy shares in Kentucky Fried Chicken, he invited his friends for dinner and tested how they enjoyed the food. 'If we try it and don't like it,' he said, 'we simply won't consider an investment in the firm.'

In 1978 Shannon became interested in the Perception Technology Corporation, which had been founded by physicist Huseyin Yilmaz from the Massachusetts Institute of Technology. Yilmaz was an expert in general relativity, but his company dealt with speech recognition. He developed a word spotter aimed at catching key words like 'bomb' or 'missile' in conversations tapped by the CIA or FBI.

Shannon knew that speech recognition was a difficult field in which research had been done for many years without much success. (In fact it would take computer experts another two decades to develop viable speech-synthesis programmes.) So he decided that Perception Technology Corporation was not a good bet.

In spite of gossip that has circulated in the scientific community for years, Shannon's investment strategy was not spectacular. He had no magic recipe for getting rich on the stock market, but had found some basic principles, which—with a grain of luck—helped him get good returns in the long run. Had the market crashed like in 1929, however, he might have been less fortunate.

Shannon never examined the reasons for the 1929 collapse. Like all investors, he was interested in his own portfolio and not in the global mechanism of the stock exchange. But the idea of picking the right data and ignoring the noise was a powerful one. Whether others discovered it independently of him or he helped it spread through word of mouth is not known. What matters is that economists adopted it.

One of those who understood what data would shed some light on the dynamics of the stock market was the Yale professor Robert Shiller. In March 2000, shortly before the Dow Jones reached a peak, he published a book in which he suggested that a downward trend was likely to come soon. The collapse came within days.

Irrational Exuberance

Shiller's book *Irrational Exuberance* became a bestseller and was translated into eighteen languages. The title was inspired from a speech Alan Greenspan, the Federal Reserve Chairman, gave in December 1996 at the annual dinner of the American Enterprise Institute for Public Policy Research in Washington DC. Talking about the price–earnings ratio and how to keep inflation down, Greenspan asked: 'But how do we know when irrational exuberance has unduly escalated asset values, which then become subject to unexpected and prolonged contractions?'

Though Greenspan's question seemed harmless, it shook the next-day markets all over the world. In Japan, the Nikkei index fell 3.2 per cent; in Hong Kong, the Hang Seng was down 2.9 per cent; in Germany, the DAX dropped 4 per cent; and in New York the Dow Jones lost 2.3 per cent soon after the Exchange opened. Shiller chose to start his book with the story of this drop. *Irrational Exuberance* answered Greenspan's question, showing that there are ways to know when the market is ready to crash.

In fact things happened the other way around. Two days ahead of Greenspan's remark, Shiller had testified before the Federal Reserve Board. The Yale professor told those present that market levels were irrationally high and outlined his arguments in this direction. Among them was the inflated value of the price–earnings ratio, which Greenspan would mention in his speech.

Irrational Exuberance developed the ideas Shiller had presented to the Federal Reserve Board. The book studied the recent stock-market boom, using published research and historical evidence. Shiller found many similarities and very few differences between the current situation and the one encountered before the crash of 1929, and concluded that history might repeat itself.

Days after Shiller's book was published, the collapse of the dotcom bubble took the market down. The Dow Jones needed six years to reach again the height of March 2000. Though Shiller could not predict the day when most investors would start selling, he showed that the premises were there. The market needed only a slight push to fall.

Ready to Burst

At the end of the twentieth century the American stock markets soared to new levels. The Dow Jones, which stood at about 3,600 in 1994, tripled in five years, exceeding 11,000 by 1999 and reaching 11,700 in the early days of the year 2000. The other economic indicators, however, showed modest growth. The personal income and gross domestic product in the United States rose less than 30 per cent, almost half of which accounted for inflation, and corporate profits did not surpass 60 per cent. These simple observations led economists to suspect that the stocks were grossly overpriced.

This state of affairs was not restricted to the United States. In Canada, France, Germany, Italy, and Spain the market doubled over the same period, while in Australia it increased by half. Indonesia, Hong Kong, Japan, Malaysia, Singapore, and South Korea as well as Brazil, Chile, and Mexico saw remarkable gains too, but none of them equalled the American surge.

Economists began to question what the future would bring. How does the current period of high stock prices compare to similar periods in the past? A steep increase is usually followed by poor performance in the following years. Does this pattern apply now too? What can be done to avoid a crash or a bear market? To answer these questions, the experts had to assess previous and current evolutions, decide if there was any danger, and act if necessary to avoid an economic breakdown.

Robert Shiller was among the economists who asked these questions. He decided to plot the American stock prices from January 1871 to January 2000 and the corresponding earnings for the same period. He thus obtained the graph in Figure 7.1.

Shiller noticed that the graph of the earnings did not grow in the same proportion as stock prices. The former lagged behind. So the profit shareholders made from keeping stocks did not justify their investments. The only way to make large profits was to resell shares at a higher price. The original purpose of the stock market—to invest in promising companies and help them grow—was thus defeated by speculation.

When investors hope that a stock price soars, they buy, and prices go up too. Logic defies them. No indicator suggesting that stocks are overpriced matters as long as there is demand for

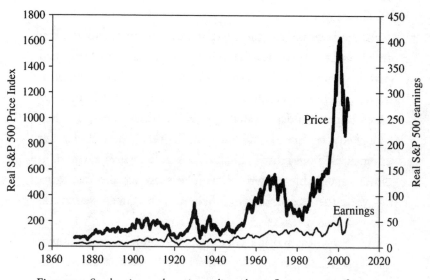

Figure 7.1 Stock prices and earnings adjusted to inflation starting from 1871.

stocks and they can make money quickly. But in the back of their minds they are warned of an existing bubble and will withdraw their money in case of danger. Until then, however, they don't want to miss the high tide.

That speculators create bubbles has always been known. Referring to speculation, the former American president Herbert Hoover wrote in his 1952 memoirs: 'There are crimes far worse than murder for which men should be reviled and punished.' But as long as trading unlimited numbers of stocks remains legal, speculation can't be curbed.

Shiller also looked at the spikes of 1929 and 2000. The former was followed by an abrupt fall. Would the latter have a similar fate? He was tempted to think so, but he didn't want to rush into conclusions. For the time being, he noticed that there had been a run-up in the 1950s and early 1960s with a flat five-year period and a drop in 1973–4. The price hike, however, had been less sharp in that case.

An explanation for the swift increase in prices in the last decade of the twentieth century was probably the unusual growth of earnings. From 1992 to 1997, they almost doubled. But the start of this trend marked the end of a recession during which prices had depreciated. Similar increases in earnings had happened after recessions in the past. Between 1921 and 1926, for instance, earnings more than quadrupled after a severe recession. Other remarkable growths took place during the five years following the recession of the 1980s and the Great Depression of the 1930s.

Shiller also plotted the price–earnings ratio—a quantity that measures how expensive the market is relative to the ability of corporations to earn profits. With the help of a ten-year average method, which evens out irrelevant information such as the surge of earnings during World War I and their brief decline during World War II, Shiller obtained the graph in Figure 7.2.

A remarkable feature of this graph is the resemblance between the spikes of September 1929, when the ratio topped at 32.6, and January 2000, when it hit 44.3. In both cases the numbers indicate the overrating of stock prices, not the increase in corporate profits. Significant spikes also occurred in 1901 and 1966, but they look modest relative to the peak of 1929, and the declines that followed were less spectacular than during the Great Depression.

There is more to learn from the graphs in Figures 7.1 and 7.2. The yearly average return of the stock market, for instance, was −13 per cent between 1929 and 1934, −1.4 per cent for the

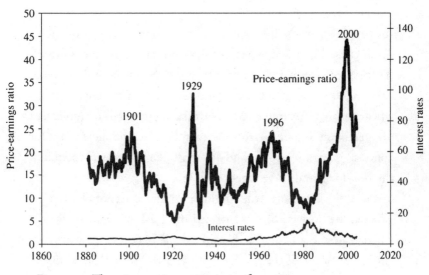

Figure 7.2 The price–earnings ratio starting from 1881.

following decade, −0.5 per cent during the next fifteen years, and +0.4 per cent for two more decades. This observation seems to shake the belief that a balanced portfolio rarely beats the average return, otherwise why would people keep investing in the stock market?

Indeed, although the Dow Jones needed more than six years to reach the peak of March 2000, not counting inflation, many investment firms did well during this period, with little or no losses in 2000 and with profits as high as 10 per cent in the following years. Those who trust the stock exchange argue therefore that the investment strategy matters, not the trend of the market. But they seem to forget the lesson of 1929, when the Great Crash ruined almost everybody.

Public Perception

In 1999 Robert Shiller surveyed a random sample of rich Americans. He asked whether they thought that stocks were the best long-term investments through the ups and downs of the market. Out of the 147 respondents, 96 per cent agreed, 2 per cent were neutral, and 2 per cent disagreed. Such high level of agreement is rare in surveys that measure public trust. In contrast, during the 1970s and 80s most people saw real estate as the best long-term investment.

To make sure that the framing of the question had not affected the answer, Shiller invoked the crash of October 1987, when in less than three weeks world markets lost between 22 and 46 per cent of their value. 'Do you think that if a similar event happens tomorrow the markets will be back to the current level in less than a couple of years?' he asked. This time 91 per cent of respondents agreed, 3 per cent were neutral, and 6 per cent disagreed. The trust in stock investments was unusually strong.

Similar optimism had existed in the 1920s. On 2 January 1925, journalist Tracy Sutliff wrote on the front page of the *New York Herald Tribune*: 'There is nothing now to be foreseen which can prevent the United States from enjoying an era of business prosperity which is entirely without equal in the pages of trade history.' She was right. The Great Crash was impossible to predict so early.

But no signs of worry appeared later in time either. In the *Atlantic Monthly* of August 1928, John Moody, the head of a rating agency, expressed similar trust in the stock market. 'A new age is taking form throughout the entire civilized world . . . We are only now beginning to realize, perhaps, that this modern, mechanistic civilization in which we live is now in the process of perfecting itself.'

In the book *Only Yesterday* by Frederick Lewis Allan, the investment mood in 1929 is captured in the following words:

The rich man's chauffeur drove with his ears laid back to catch the news of an impending move in Bethlehem Steel; he held fifty shares himself on a twenty-point margin. The window cleaner at the broker's office paused to watch the ticker, for he was thinking of converting his laboriously accumulated savings into a few shares of Simmons. . . . [It is also] told of a broker's valet who made nearly a quarter of a million in the market; of a trained nurse who cleaned up thirty thousand following the tips given her by grateful patients; of a Wyoming cattleman, thirty miles from the nearest railroad, who bought or sold a thousand shares a day.

Even some experts thought that this prosperous period would never end. Irving Fisher, a Yale professor, was quoted as saying just before the 1929 peak that 'stock prices have reached what looks like a permanently high plateau'. Craig B. Hazelwood, a bank's president, went as far as to claim in the 2 January 1929 issue of the *New York Herald Tribune* that prohibition was among the guarantees of the market's stability: 'Most of the money formerly spent in the saloon has since gone into continually higher standards of living, investments and bank savings.'

This display of optimism was not unjustified. The economic growth of the 1920s had been remarkable. Electrification was extended beyond the major cities, and the radio became a household item. In 1920 there were three American radio stations; three years later, more than 500. Sound also came to projection theatres in 1923, eliminating the silent movie by the end of the decade. If in 1914 there had been only 1.7 million cars registered in the United States, their number grew to 8.1 million by 1920 and to 23.1 million by 1929. Most areas of the economy developed at a similar pace.

It would be misleading, however, to leave the impression that nobody rang the bells of alarm. The high price of stocks worried a few experts. In the *Commercial and Financial Chronicle* of 9 March 1929, for instance, Paul Warburg of the International Acceptance Bank warned about the dangerous level of market speculation. In the *New York Times*, several economists drew attention to the perils of the current market trend. But investors didn't listen. They kept buying stocks at record prices.

The 1920s are not the only attested case of irrational exuberance. Perhaps the oldest significant example is that of the tulip bulbs, which reached astronomical prices in the first half of the seventeenth century. The term *tulipomania* or *tulip mania*, which now stands for economic bubbles, is derived from that experience.

Introduced to Europe from Turkey in the mid sixteenth century, the tulip became popular among the upper classes of Dutch society. Some rare varieties turned into a status symbol. As a consequence, the price of the tulip bulb surged. In 1623, a single bulb of a famous tulip variety was sold for almost 1,000 florins, or seven times the yearly average income. But this was not the upper limit. In 1635 an investor offered 6,000 florins for one bulb of the *Semper Augustus* tulip.

In February 1637, however, prices reached a plateau. Many speculators who had become rich could not make money anymore. As a result, prices fell. Within weeks, they plummeted ten times, ruining most business people involved in the trade.

A similar exuberance gripped the investors of the 1990s. The surge of the computer industry and the Internet marked the beginning of a new era. Like seven decades earlier, the economy was booming. The 1929 crash seemed too remote to matter. Investment analysts found reasons to think that the current situation was different, the price of stocks reflected a new reality, and the market was safe. The situation was different indeed, for it is never exactly the same as in the past, but they failed to explain why this difference guarantees security.

On 30 December 1996, Michael Mandel, economics editor at *Business Week*, published an article in which he gave several reasons for the market's health. They were: the surge of the high-tech industry, growing profits, falling interest rates, increased globalization, and low inflation. While the merit of each factor is debatable, the last one is particularly controversial. In the 1960s, analysts agreed that high inflation would push the market up. Now they thought the opposite.

In truth, most articles published in the business press are marked with wishful thinking. Investors have therefore a vague, often distorted view about the past, and are burdened with unrealistic beliefs. Many think the 1929 crash happened in one day. Few know that the market recovered after the Black Monday, but fell for the next three years. And almost everybody trusts that the market will always go up in the long run. A careful look at Figure 7.1, however, contradicts these beliefs.

So do markets always drop after a growth period? Is every 'new era' doomed in spite of attempts to justify that things will turn out better this time? Many have asked these questions, but few have found good answers.

Complex Systems

One of those who approached the problem of stock market crashes was Didier Sornette, a former professor of geophysics at the University of California in Los Angeles who now holds the Chair of Entrepreneurial Risks at the Swiss Polytechnic Institute in Zurich. His original interests had been in earthquakes and ruptures, which apparently have nothing in common with economic breakdowns. (We met him briefly in Chapter 2 in connection with the *Nature* debates on earthquake prediction.) But after he studied the theory of complex systems, Sornette realized that earthquakes and financial crashes are two sides of the same coin.

He disagreed with Alan Greenspan, the President of the Federal Reserve Board, who thought it 'very difficult to definitively identify a bubble until after the fact'. As a scientist, Sornette believed in the possibility of predictions. He relied on the theory of complex systems—an interdisciplinary field born from the desire to understand time-evolving entities with multiple interactions between their components. Some typical examples of complex systems are cells, living beings, human economies, and transportation networks. Though non-linear and chaotic in their evolution, they are endowed with memory and feedback, which help them adapt to new conditions.

Most complex systems exhibit rare and sudden transitions called extreme events or catastrophes. An earthquake, for instance, is a catastrophe for the complex system of geologic plates, and so is a financial crash for the stock exchange. An important goal in the field of complex systems is to predict extreme events.

Scientists have assumed for a long time that catastrophes are nothing but large-scale versions of small changes: a major earthquake is a significant tremor of the Earth's crust, and a market crash is a magnified price drop at the stock exchange. In other words, size is the only difference between extreme and usual events.

The theory of complex systems challenges this view. A catastrophe is regarded as distinct. Its circumstances are different from the ones that produce common changes. A large earthquake, for instance, is the result of some amplifying mechanisms, which are absent in small earthquakes, and a financial crash occurs because of developments not seen for an ordinary price drop. The identification of the specific factors that trigger extreme events becomes essential for prediction.

Didier Sornette adopted this novel point of view when he started thinking about financial collapses. The most recent literature gave various reasons for market crashes, including computer and off-hours trading, derivative securities, lack of liquidity, trade and budget deficits, overvaluation of stocks, and many more. But he found them unsatisfactory. They had been around even when stock prices went up.

Sornette hoped to find a better clue in a book of research papers published in 1989 by the Mid American Institute for Public Policy Research. The authors of an article had looked at a statistical variable that explained the October 1987 crash in terms of the local stock-market response. More precisely, they introduced a world market index that involved the currencies of twenty-three major industrial countries. If on 30 September 1987 this index was given the value 100, it fell to 73.6 a month later.

The world index was statistically related to monthly returns in every country from early 1981 until September 1987. This correlation signalled an involuntary worldwide cooperation between investors before the crash. In other words, the crash in each country was explained as a response to a worldwide crash. But what triggered the latter, an overlap of all national crashes?

This question reminded Sornette of the chicken-and-egg problem. The world index was not the indicator he was looking for. Still, it suggested something he had expected all along. The market is not efficient and unpredictable, as believed, but shows a self-organized structure in which cooperation, imitation, herding, and feedback play a key role in the development of instabilities. As it grows, the bubble becomes less robust and more susceptible to internal disturbances. Its collapse lies therefore in the unsustainable inflation of stock prices.

When blowing air into a rubber balloon, we can identify with some approximation the limit beyond which the balloon bursts. We develop an intuition for how tight the rubber gets when the air pressure nears the level that exceeds the balloon's resistance. A scientist who does some measurements could tell exactly when the blast occurs. But are there indicators that predict when a stock-market bubble bursts? Sornette believed there were, and he wanted to find them.

Bubbles and Anti bubbles

A basic principle in the theory of complex systems is to look at the forest rather than the trees. Indeed, the local details get complicated quickly and predictions become difficult to make. The global picture, however, may offer some insight into the future. The motion of plankton, for instance, is hard to foresee by studying the drift of its many individuals. But the study of the entire group from the perspective of the surrounding environment might lead to a meaningful forecast.

The complex system of the stock market withstands external intervention because investors react to internal changes alone. If the market goes up, and everybody makes money, people won't sell their stocks for safer but less profitable assets. In the absence of downturn signs, they keep buying at record prices, thus stretching the bubble even more.

Sornette realized that the entire market leads to the premises of a crash when the system reaches a high-enough instability, whose level can be measured through a global study of stock prices. To perform this analysis, he resorted to mathematics. His main tool was the theory of fractals.

The Yale mathematician Benoit Mandelbrot had coined the term *fractal* in 1975, following certain developments in the theory of dynamical systems. Intuitively, fractals are geometric shapes that can be divided into parts resembling the whole. In other words, every zoom in reveals a similar image.

Fractals allowed mathematicians to generalize the idea of space dimension to fractions. The ancient Greeks had figured out that the line has one dimension, the plane two, and the ambient space three. But nobody had envisioned a geometrical entity of, say, two-and-a-half dimensions, which some fractals have because of their structure. Of course, fractal dimension is not an intuitive notion. We cannot imagine it—as we cannot visualize a ten-dimensional space—but mathematicians have learned how to deal with this concept.

Sornette's graphs, which represent the variation of prices on the stock market, look like fractals. Indeed, every portion of the graph in Figure 7.1 is composed of many stock prices. The price of each stock results from the trade between thousands of agents, and so on. So every zoom in on the graph leads to information that resembles the global one.

Finally, Sornette applied some mathematical techniques to discard the irrelevant information from the studied graphs. This operation resembled Shannon's idea of stripping the data from noise and understanding the basic tendency of the market by ignoring the local fluctuations. After performing many laborious calculations, Sornette found out whether a build-up towards an extreme event occurred.

Furthermore, he did empirical tests to see whether his methods revealed price patterns characteristic to pre-crash periods. Such 'fingerprints' appeared indeed before the main historical market collapses, but he could not identify them in the current stock markets around the world, so he expected no crash in the near future.

He also defined the concept of *anti-bubble* as a low in a market when stocks are undervalued and prices can only increase. The analysis he made with his postdoctoral scholar Anders Johansen showed anti-bubble patterns in Japan. In January 1999 they predicted a 50 per cent increase of the Nikkei index by the end of the year although Japan's economic indicators were declining and market analysts had just warned of a further stock-price drop. By December, the Nikkei index was up 49 per cent.

Sornette also made successful predictions for several short-term changes of trends in the United States and Japan. Nevertheless, he admitted that his results are never 100 per cent accurate because the stock market behaves more like the weather than like a balloon whose burst can be foreseen correctly. Sornette can forecast tendencies on the stock exchange, and hopes to do better when his methods grow more refined.

In September 2002, for instance, he predicted that the Standard and Poor's 500 Index would continue its uptrend for no more than a few months and then descend about 20 per cent into the first semester of 2004. In reality, from a high of 924 in September, the index reached a low of 768 in October. After a slight increase in the following months, it hit another low of 788 in March 2003. From then on until the end of March 2004, the index rose slightly on the average until it levelled at about 1100. So though this prediction was not fully accurate, Sornette had captured the indicator's general tendency.

Real-Estate Crashes

Many people are preoccupied with house prices. Is the real estate a bubble? Did we pay too much for our house? When should we sell our property? After all, these concerns are justified. For most of us, the house in which we live is *the* investment of our lifetime. Therefore Sornette also wanted to see if his methodology could discover when real-estate markets are overpriced.

After the crash of the year 2000, the Federal Reserve reduced short-term rates from 6.5 to 1 per cent in less than two years. The change was necessary in the context of a strong housing demand combined with a weakening of the American economy. This approach was successful. People invested in real estate and the economy recovered. But soon house prices reached new heights.

In March 2003, Sornette and his postdoctoral fellow Wei-Xing Zhou released a paper in which they claimed that this trend could lead to a new real-estate bubble in the United States. Britain's market seemed to have already built such a bubble. Their prediction: If rates won't change and the demand for mortgages continues, two turning points will occur in Britain: one at the end of 2003 and the other in mid 2004.

Though some local changes did occur on the British market, with prices falling in some months relative to the previous one, no crash occurred. In 2008, however, prices started to go down. Sornette and Zhou argued that their prediction signalled only a change of regime, which could mean stabilization or a slight decrease in prices, but not necessarily a crash. Indeed, real estate rarely crashes because, if having a choice, owners don't sell below a certain price. Therefore what usually signals the change of regime is the volume of sale, which decreased indeed as the two experts had predicted.

In May 2005, Sornette and Zhou released another paper. They provided evidence of an American real-estate market that grows faster than exponentially. This trend signalled a bubble, which—they predicted—was likely to end in mid 2006. Again, though no immediate crash occurred, the House Price Index growing 0.86 per cent in the third quarter of 2006 relative to the second quarter, the volume of sales agreed with the prediction, and a year later, house prices started to drop. At the time I am finishing this chapter (October 2008) they went down dramatically, proving Sornette and Zhou right.

The reason why predictions are not always correct is that outside interventions occur from the time a forecast is released. For instance, in 2006 the Federal Reserve increased interest rates. From the historically low rate of 1 per cent, the Board applied quarter-of-a-point increments at most of its successive meetings, eight of which take place every year, bringing the rate to 5.25 per cent in mid 2006. Gradually, more and more people thought twice before investing in real estate, and the bubble's growth was contained for a while. But indiscriminate lending practices led to a credit crunch that lowered house prices.

This crunch created a sharp drop in stock markets, and at the time this manuscript has gone into production we still don't know the outcome of the crisis. The large bailout of financial institutions in the United States and the nationalization of some banks in Britain, followed by similar steps all over the world, seem to have stopped the disaster for now. But the difficult times are not over, and nobody knows how these interventions will influence investors' behaviour. This state of confusion, with daily ups and downs never seen before, is well captured in the words of Paul Krugman, the 2008 winner of the Nobel Prize for Economics, who declared soon after the British rescue plan was announced: 'I'm slightly less terrified today than I was on Friday.'

Of course, no mathematical model can predict a sequence of outside interventions, which change the original problem. Therefore Sornette makes no claim that he can always predict crashes. What his pattern-recognition technique does is forecast the end of a bubble. This is the most likely moment of the crash, though it can happen sometime before or after that critical instant, or not at all. The bubble could deflate slowly over a long period or stabilize in time. In the case of the current financial crisis, he predicted the real-estate bubble correctly.

Predictions of this kind become most efficient when they can convince decision makers to act against a crash. The goal of the astronomers who seek Near-Earth Objects is not to see them collide with Earth. Rather, they want to signal the danger and force us to take preventive action. Similarly, the collective purpose of market-crash forecasts is to avoid a financial disaster that would wipe out our savings, bankrupt our pension plans, push prices and unemployment rates up, and disrupt our social and economic milieu.

The theory of complex systems has proved suitable in the study of the stock market. Other mathematical techniques are also used to discover signs of economic danger, whether in stock trade, real-estate transactions, or inflation tendencies. A lot of progress has been made in the past decades for understanding these complicated phenomena towards building a safer economic world. But new situations show up, such as the current crisis, which has similarities but is not identical with either of the 1929 and 1873 drops, the latter leading to a depression that lasted six years. Since experts often find themselves in uncharted territory, the prediction and prevention of stock-market crashes is likely to remain a big challenge for many years to come.

8. TINY KILLERS: PANDEMICS

When we think of the major threats to our national security, the first to come to mind are nuclear proliferation, rogue states and global terrorism. But another kind of threat lurks beyond our shores, one from nature, not humans—an avian flu pandemic.

Barack Obama

Among the many evils I encountered while growing up in Romania during Ceauşescu's dictatorship were some good things too. One of them was the educational system. Although the authoritarian school structure was aimed at keeping students busy, specialized teachers taught each subject starting with grade five. So I always had mathematics graduates in my mathematics classes, philology graduates in literature, and arts graduates in drawing and painting.

My mother was the only biology teacher at school, and—whether we liked it or not—I became her student. A memory I have from that time is my mother's likening of the white blood cells to soldiers—a classic metaphor that is still used today. 'When a virus invades,' she told me, 'the army surrounds it. The war that follows makes us feel tired and sick. After days of fierce battle, the fever recedes, the virus is defeated, and our body starts recovering from the ordeal. This period is critical because the enemy can strike again. Therefore we must keep

taking the prescribed medication, which plays the role of rein-
forcements in our army's depleted ranks.'

I remembered this metaphor whenever I fell ill with the flu,
so I never ignored the physician's advice regarding the length of
the treatment. More recently, this memory enticed me to learn
more about how our body reacts to disease. The details are more
complicated than my mother had suggested, therefore it took
me a while to understand them. But the connections I grasped
are fascinating, and they made me wonder whether pandemics
can be predicted. This question led me back to mathematics and
the models many biologists use.

The Plague

Although pandemics have always haunted the world, their
origins are not easy to trace in the distant past. All we know is
the description of some symptoms and the (possibly exaggerated) death toll the diseases brought, but we don't know what
produced them. Since medical knowledge was poor, and little
distinction was made between various epidemics, the ailments
that killed legions of people during antiquity and the Middle
Ages are known under a single name: plague.

The earliest account that describes such an epidemic appears
in the Torah, which tells that a plague hit the Philistines of
Ashdod for stealing the Ark of the Covenant from the Children
of Israel. In his *History of the Peloponnesian War*, Thucydides
mentions the pandemic of 430 BC, which started in Ethiopia,
extended to Egypt and Libya, and reached the Greek city states,
killing about one out of three Athenians, including their king,

Pericles. The disease struck Thucydides too, but he was lucky to recover.

In the first century AD, the Greek anatomist Rufus of Ephesus mentions an outbreak, which decimated the population of Libya, Egypt, and Syria. The years 541 and 542 are infamous for a plague that struck the Eastern Roman Empire during the reign of Emperor Justinian I. At its peak, the epidemic allegedly killed 10,000 individuals a day in Constantinople alone. With almost one quarter of the east Mediterranean people wiped out, nobody escaped the consequences of the pandemic. But these numbers were eclipsed in 588 when an estimated twenty-five million people died in south-west Europe.

The Middle Ages experienced even larger waves of destruction. Between 1347 and 1350, the 'Black Death' changed the course of history by killing more than one third of the people in Asia and Europe. Local outbreaks continued to occur in the next three centuries with diminished intensity. One of them, which hit London and its vicinity in 1665, is famous in the history of science: Isaac Newton fled Cambridge for his Woolsthorpe farm in Lincolnshire, where he invented calculus and discovered the law of gravity during his *annus mirabilis*—the wonder year.

A last big outbreak struck Asia in the late nineteenth and early twentieth century. By that time, the French scientist Alexandre Yersin and the Japanese Shibasaburo Kitasato isolated the bacterium responsible for what we now identify as 'the plague' and explained that it was spread by fleas who bit infected rodents, thus offering a clue of what should be done to avoid future pandemics. So far, the prevention measures have been successful, with only isolated outbreaks reported in the past few years.

If bacterial infections like those produced by the plague can be contained through careful prevention procedures, it is more difficult to stop the spread of viral epidemics. The Spanish flu that hit the world in the early twentieth century provides such an example.

Influenza, 1918

At the beginning of 1918 there were clear signs that World War I was coming to an end. More than nineteen million people had died, almost half of them civilians. The planet was eager to embrace a lasting peace. But nobody noticed the even larger disaster looming in the distance: the deadliest influenza pandemic ever seen.

It erupted in pockets around the world without any indication that it would be more than a seasonal cold. Today, some researchers claim it was observed as early as January, but most agree that it had definitely appeared in March. Yet, it took a while to spread and become recognized as a problem of utmost proportions. The armies involved in the war contributed to the delayed response because none of them wanted to admit casualties due to a rampant disease.

The symptoms included an obstruction of the lungs that led to coughing blood. Often, the flu acted quickly. People could be suddenly struck and die the next day. In many cases patients drowned in their own body fluids. Some less violent cases led to pneumonia with neural complications, and many of those who survived this form of disease suffered mental disorders for the rest of their lives. The pandemic also killed many adults aged

20 to 40, unlike the common flu, which usually hits young children and elderly people.

This influenza became known as the Spanish flu not for starting in the Iberian Peninsula but because it received more press coverage in Spain, which—as a neutral country during the war—did not censor the news related to the disease. The virus killed more than fifty million people worldwide, half of them in the first twenty-five weeks of evolution. (In comparison, AIDS saw twenty-five million victims in its first twenty-five years.) About one in five people were infected, and the global mortality rate was estimated somewhere between 2 and 2.5 per cent, much higher than the usual 0.1 per cent registered for the common flu.

Numbers, however, varied from region to region. Some communities were wiped out, while others suffered insignificant losses. In the United States, which had more than half a million victims—ten times the number of Americans killed in the war—the average life span fell by ten years. In India, one in every twenty citizens, or seventeen million people, died. Japan, with just over a quarter of a million deaths, had a mortality rate of only 0.425 per cent, by far the lowest in Asia. There also existed unaffected areas, such as the Marajó Island in Brazil and the French colony of New Caledonia, but they were exceptions, which showed how effective isolation and quarantine could be when properly enforced.

The origins of the Spanish flu were unknown for a long time. Some speculated that it had come from China in a rare genetic shift of the influenza virus. Others thought it was a result of the war: anywhere from a biological weapon produced by the Germans to the unwanted side-effect of mustard gas. More credible, however, was that the virus had travelled from poultry to humans through swine. But some recent evidence favours a direct link between poultry and humans—the main reason why fear of the avian flu has reached cosmic proportions in recent years.

This new insight came after a team of the US Armed Forces Institute of Pathology obtained samples of the Spanish flu virus in 1998 from the frozen body of an Inuit woman who had died eighty years earlier in Alaska. Several experts analysed the genetic structure of the virus and compared it with the only two other existing laboratory samples. In 2004, John Skehel, director of the National Institute for Medical Research in London and Ian Wilson of the Scripps Research Institute in San Diego synthesized a crucial protein from the virus, whose genetic sequence would be described a year later. This result proved the direct link with the virus responsible for the avian flu.

The key to predicting the possibility of a new epidemic is to understand if a particular virus carried by certain animals can mutate, get transmitted, and finally spread among humans. But what are the mutations that must take place to enact such a scenario?

Of Birds and Men

In the past few years, millions of birds have been slaughtered for the fear of an avian flu pandemic. In China, the virus killed several people exposed to poultry in 1997, but no human-to-human transmission occurred. Nevertheless, the experts feared that the virus could mutate and lead to a dangerous epidemic. Still, nobody knew what kind of mutations would produce a potential disaster. Therefore many research institutes around the globe began to look into this problem.

A breakthrough occurred in November 2006, when an international research team published its findings in the journal *Nature*. Yoshihiro Kawaoka, a virologist at the University of Tokyo and the lead author of the paper, pointed out that the team had identified two important mutations the avian flu virus H5N1 had made towards recognizing human receptors.

Receptors are molecules that allow the virus to penetrate the cell and initiate an infection. Birds and humans have different receptors, so normally avian viruses cannot enter human cells. But such viruses evolve quickly by exchanging genetic information. Apparently, when the avian flu virus and the common human influenza virus meet, they swap genes. So humans exposed to both viruses at the same time give the avian flu virus the possibility to alter its genetic structure.

The viruses Kawaoka's team found in ducks and chickens didn't recognize human receptors. The bad news was that they had already mutated since 1997, when first discovered, in the direction of penetrating the human cell. Besides the two changes the team identified, the avian flu virus needs some other adjustments to spread among humans. So could H5N1 adapt and produce an epidemic in the future? A look at the past might give us a hint.

Since 1918, at least two other viruses mutated to recognize human receptors. Both triggered significant pandemics. The first, catalogued as a virus of the H2N2 subtype, which led to the so-called Asian flu, spread worldwide in 1957–8; the second, H3N2, responsible for the Hong-Kong flu, hit in 1968–9. Though not as severe as the Spanish influenza, these pandemics killed about three million people together. So, by analogy, we may be tempted to answer the above question with: Yes, H5N1 could adapt to spread among humans and produce a pandemic.

But a closer look at the history of epidemic diseases shows that some would-be pandemics never happened. A famous case is that of the 1976 swine-flu virus, which scared the American authorities into ordering mass immunization. After more than forty million people had received vaccines, the programme was suspended when the experts understood that the virus was far less dangerous than thought.

Another interesting case is that of the Russian flu, which began in May 1977 with a virus spreading rapidly in children and young adults. Those born before 1957 were only marginally affected because the virus had also circulated before that date, thus immunizing most people. The Russian flu is therefore referred to as a 'benign pandemic'.

Other alerts occurred in the recent past, none of which led to a global problem. Among them was the avian virus H7N7 in 2003, when eighty-three people fell ill and one died, and the avian virus H5N1 in 2004, mentioned earlier, which led to several deaths. The killing of millions of poultry stopped the spread of this disease in certain zones of the globe.

Computer Simulations

Viruses from animal species have the potential to mutate and allow human-to-human transmission, thus opening the possibility of a pandemic. But is there a way to learn in advance which virus will mutate in a dangerous direction? Such knowledge would be crucial for preventing the spread of the disease or minimizing its consequences.

Mathematical models offer some insight when dealing with such questions. The problem with viruses, however, is that models that accurately predict how new subtypes will behave are difficult to construct based on the known information. Sufficient data can usually be gathered only after the fact. To this category of models belong those based on differential equations. Though the most accurate, they are of little use today in predicting a future pandemic. Nevertheless, they play key roles for understanding the spread of diseases, as we will see later.

But mathematics has many faces, and one of them proved useful to Neil Ferguson, a professor of mathematical biology at the Imperial College in London. Together with Alison Galvani and Robin Bush from the University of California, he studied the genetic variation of new influenza strains with the help of *stochastic models*, which offer probabilities of future events. In their 2003 article published in the journal *Nature*, they kept track of individuals exposed to influenza, and found the probabilities that some population segments get infected. To obtain these results, they took into account various parameters, such as the genetic information of the virus, how likely it was to mutate, and whether part of the population was immune due to previous exposure to similar strains.

Their techniques involved computer simulations, which helped them perform lengthy calculations that would have been impossible to do by hand. Though still in an early stage of development, such models might provide in a few years concrete results in specific cases. They may tell, for instance, that a certain influenza strain is likely to develop pandemic-like characteristics, thus giving the medical authorities some time to react with preventive measures.

Thus far, however, neither Ferguson's research nor any other achievements in the field have succeeded in making a deterministic prediction regarding the development of a deadly epidemic. Nevertheless, the existing results can help alleviate the consequences of an incoming pandemic. The state of the art in forecasting such events is therefore similar to the one we encountered in the chapter on tsunamis, where a warning at short notice can save many lives.

In this research direction, Ferguson and several of his colleagues have recently become interested in strategies for mitigating influenza pandemics. They have used some complicated mathematical models based on differential equations to determine the contribution of each possible action that can be taken to diminish the evolution of the disease.

The experts divide prevention and containment strategies into three large categories: antiviral, vaccines, and non-pharmaceutical. The first strategy refers to the specific medication used once a disease occurs, the second to preventing infection through immunization, and the third to diminishing the spread of the virus by travel restrictions, workplace and school closure, isolation, and quarantine. Ferguson and his colleagues wanted to learn which combination of measures is more effective and how to implement it after a flu outbreak.

Models need assumptions, and the authors restricted their study to Britain and the United States, which are rich in data. This information allowed them to assume that in the United States, for instance, 30 per cent of transmissions occur within the household and 70 per cent outside it, the latter category divided into 33 per cent of infections occurring in the general community and 37 per cent at school or work. Other hypotheses concerned how infectious the virus is and what the transmission scenarios are.

Some of the team's conclusions were unexpected. Border and travel restrictions, for instance, only delayed the spread of the disease by two or three weeks, unless they were at least 99 per cent effective. Though 90 per cent efficacy still had some impact, 75 per cent had almost none. School and workplace closure applied at the peak of the outbreak could reduce infection during short periods by as much as 40 per cent, but had little influence over the general rates of spread. Household quarantine and isolation, however, showed significant effects on stopping the infection.

The study also confirmed well-known facts. For example, drugs were effective, but only if the antiviral medication was administered within a day after the symptoms occurred. Similarly, prior vaccination proved helpful against the spread of the disease even when the vaccine was not a perfect match for the virus.

Models of this kind allow the evaluation of various scenarios by changing the input data. So once the model exists, predicting the best strategies for coping with the disease requires only realistic information about the history of transmission in a certain region or country. For individuals, the best action is to stay home during the peak of the pandemic and avoid contact with anybody outside. Of course, such a self-imposed quarantine requires enough food supplies and financial resources to skip work for several weeks, a choice most people cannot afford.

These results are among the many contributions obtained with the help of mathematical models towards predicting and preventing large epidemics. But models alone are not enough to find the best ways to proceed when a pandemic is about to engulf the globe. The experts need to look at this problem from many angles.

To get an idea of what various points of view can teach us, consider a recent remark made by Stephen Morse from the Department of Epidemiology and Centre for Public Health Preparedness at Columbia University: 'Escaping the pandemic entirely,' he wrote, 'may not be the most advantageous strategy because it only leaves everyone susceptible to infection later.' Indeed, the example of the native North Americans, a large percentage of which died of smallpox when the white man spread the virus on the continent, is an example of what can happen to a population exposed to a new disease. We can therefore learn something by looking into the past.

The Lessons of History

The *Andromeda Strain* by Michael Crichton describes how a cosmic micro-organism kills its victims in minutes by clogging their arteries. In the end the micro-organism mutates due to its contact with the human body and becomes a harmless parasite. Though fictitious, this story teaches us to keep an open mind about the evolution of epidemic diseases. A review of the past shows that the number of victims may differ from place to place even within the same country. The 1918 pandemic, for instance, registered far fewer deaths in St Louis than in other parts of the United States. How was this anomaly possible?

In 1918, each American city fought the pandemic in a different way. The non-pharmaceutical response varied to similar medical resources. The options were very much like today, from wearing surgical masks to closing schools, churches, and theatres, banning public gatherings, and isolating infected individuals. But results like those of Ferguson and his colleagues didn't exist, so each community tailored its own approach. The data of the outcomes provide a rich resource for designing future strategies.

One thing we learn from history is that timing was critical. Measures were far more effective when applied early. Communities that imposed several restrictions from the beginning (before 0.2 to 0.3 per cent of the population died) had peak death rates approximately half of those registered where the interventions started later.

An equally important measure was the maintenance of the restrictions for as long as possible. Few cities kept them in place for longer than six weeks, which was about the time it took the epidemic to pass through the area. Measures were then relaxed, and a new wave rippled the community. The lowest death rates occurred in those places that kept the control measures enforced even after the situation improved.

It is difficult to enforce social distancing for too long. Insecurity, deprivation, and psychological fatigue wear people down, and the frustration that ensues can lead to a revolt. In 1918, San Francisco saw demonstrations in which defiant citizens removed their own surgical masks. Such reactions made sociologists ask whether today, in an era of more western liberalism, restrictions would be harder to impose. Some experts dismiss the difficulties, citing as an example the Greater Toronto Area, where the Canadian public complied with the control measures enforced during the recent epidemic threat of the *severe acute respiratory syndrome* (SARS).

But new problems occur today. Globalization of travel could lead to a worldwide spread of an infection within days. The unequal distribution of medical resources puts the poor countries in a difficult position. Without collaboration, humankind might see huge social and political problems resulting from a new pandemic. The tensions arising from such a perspective are already felt. Indonesia, for instance, recently objected to providing new virus samples to the World Health Organization unless it received some assurance that it would be given foreign assistance if an outbreak occurred.

Though almost a century has passed since 1918, and two more viral pandemics haunted the world, we have learned little from the past. But a new trend is developing today. Some scientists use the historical record to predict the spread of a future pandemic or to avoid it altogether through immunization. Some experts even suggested a name for this new branch of science: 'clioepidemiology', after Clio—the muse of history. Research in this direction, however, is in its early stages.

Travelling One-Dollar Bills

If pandemics developed slowly in the past and lasted for one or two years, the dynamics of transmittable diseases have radically changed. Travel is more intense than ever, with 10,000 daily regular airplane flights in the United States alone. Under such circumstances, it is important to know how a virus would spread.

Tracking the travel habits of a nation for this purpose is not only difficult but also unnecessary. After all, only a fraction of those who travel carry viruses, and only some of those who get into contact with carriers get infected. Probabilistic models can help, but only in general terms. Though they estimate how many people might fall ill, they don't give the precise path the virus follows from one individual to another. So can we find out, in the absence of disease, how a potential virus would spread?

Three scientists asked this question and came up with an ingenious solution. Theo Geisel, the director of the Max Planck Institute for Dynamics and Self-Organization and a professor at the University of Göttingen in Germany, his postdoctoral fellow Dirk Brockmann, and Lars Hufnagel of the University of California at Santa Barbara published their results in 2006 in the journal *Nature*. Their idea was to track the movements of some half a million one-dollar bills in the United States. Like viruses, money is exchanged between some, but not all, individuals who make contact. Therefore the dispersal of bills can be used as a substitute for the spread of a virus.

These researchers used an Internet game found at www.wheresgeorge.com, played by entering the denomination, series, and serial number of any American dollar bill. What they found was not too surprising: the money moved chaotically both at local and global levels. But it was interesting that they could characterize the dispersion in terms of what mathematicians call 'Lévy flights', named after the French mathematician Paul Pierre Lévy (1886–1971). These movements are characteristic of random walks with many short steps mixed with rare long jumps. In other words, the bills changed many hands in the same city before showing up in some other part of the United States.

There is a close connection between Lévy flights and Mandelbrot sets, mentioned in the previous chapter in connection with predicting the chaotic evolution of the stock market. The unifying language of mathematics discovers relationships between phenomena that have apparently nothing in common.

The three researchers also wanted to know how far their results agreed with general human movement. Data published by the US Bureau of Transportation Statistics on the travel patterns of the American population showed them to match. This was good news, because it strengthened the assumption that the dispersion of bills is a reasonable substitute for the spread of a virus.

Still, a question they didn't ask was whether movement patterns change when a pandemic appears. Indeed, when the recent SARS threat occurred in Canada, many Canadians refrained from travelling to China, where the disease had originated. Nevertheless, the results of the three experts are still valid during the first part of an epidemic, until the authorities suggest or impose travel restrictions, isolation, and quarantine. Such decisions are crucial when predicting the spread of a disease.

A Differentiated Approach

Gaps in science get filled sooner or later. Once a paper is published, scientists read it, think of the questions that arise from there, and try to answer them. So it is no wonder that Richard Larson, a professor at the Massachusetts Institute of Technology and a former president of the Institute for

Operations Research and the Management Sciences, recently considered the issue of influenza spreading within a heterogeneous population, thus extending the work of the researchers who had studied the movements of the one-dollar bills. His point was that the forecast of the disease's evolution depends on the susceptible individuals.

Larson argued that the virus spreads less in populations with some degree of immunity than in those never exposed; that outgoing individuals, who make contact with many people, are more likely to get ill compared to those who prefer the privacy of their home; and that measures of social distancing can help during an outbreak. 'We allow [in our model] for socially active people who interact with many other people on a given day, and we allow for relatively inactive people who interact with few others,' he wrote in a paper published in the spring of 2007 in the journal *Operations Research*. 'We provide for highly susceptible people who are more likely to become infected once exposed to the virus, and we include those who are less susceptible. In a similar vein, we allow for highly contagious as well as less contagious infected persons.'

Larson's models thus capture these features, aiming to untangle the evolution of the disease and find ways to limit its negative consequences. The computations led him to two main conclusions. First, susceptible individuals who have a lot of contact with others may dominate the early exponential growth of the disease, a characteristic that might be of little use for predicting the risk of infection in the general population. And second, social distancing, combined with suitable hygienic steps, is probably an effective non-medical measure to limit and, hopefully, stop the spread of the disease.

Larson's findings agree with some of the results mentioned earlier, and they are also supported by data collected in the field. The recent SARS threat in the Greater Toronto Area in Canada provides such an example.

Isolation and Quarantine

On 23 February 2003, a woman who returned to Toronto from Hong Kong was diagnosed with SARS. About 250 people living in the area got infected with the virus in the following weeks, and forty-three of them died. Meanwhile, the medical authority of Ontario sounded the alarm and implemented restrictions, which included tough medical-control practices in the hospitals of and near Toronto. No cases of infection occurred after 20 April, so ten days later the World Health Organization lifted a travel advisory issued earlier. It looked like the isolation and quarantine measures succeeded in stopping the spread of the disease, which could have easily developed into a global pandemic.

A similar case occurred in Vancouver at about the same time, but the victim sought medical assistance right away. The hospital personnel acted quickly and efficiently, and though several probable cases of SARS were reported, nobody died.

These events prompted three researchers to develop a model that could predict how efficient isolation and quarantine are in fighting SARS. The work initiated by Carlos Castillo-Chavez, now at Arizona State University, Carlos Castillo-Garsow of the State University of New York at Buffalo, and Abdul-Aziz Yakubu of Howard University in Washington DC was not a mere after-the-fact analysis. These experts aimed to understand whether the measures taken in Vancouver and Toronto would always contain the disease or might have unexpected consequences under different scenarios.

They regarded this problem from a wide perspective. Isolation and quarantine measures are not appropriate for all infectious diseases. Chicken pox, for instance, has mild effects but requires long periods of isolation. The high cost of missed work would be thus hard to justify. Other solutions, such as vaccination of the entire child population, are more effective in such circumstances.

There are also cases when isolation and quarantine can be harmful in the long run. Using this measure to control rubella in China, for instance, might increase the level of disease. Indeed, under the current regulations of no control, 97 per cent of the population is immune to the virus because these people formed antibodies while exposed to infected individuals. Isolation and quarantine would make more people susceptible and might consequently make things worse.

The model the three experts developed was limited to the duration of a single SARS outbreak. This condition simplified their work but also made it less realistic. Nevertheless, their assumption did not defeat the original goal, which was to understand the effectiveness of isolation and quarantine under general circumstances when a SARS infection occurs.

The main reason why the team decided to keep the model simple was that the addition of more realistic elements runs the risk of imposing unrealistic restrictions on them. Indeed, many of the interactions that occur in such models depend on human decisions, which are difficult to quantify. It is unrealistic to say that everybody will react in a certain way to a given stimulus. Since this model was a dynamical one that did not involve probabilistic elements, it could not capture the whims of social behaviour. Still, it was worth pursuing for understanding the effectiveness of isolation and quarantine.

As in most models of this type, the variables included several groups of individuals: susceptible, exposed, infectious but undiagnosed, infectious and diagnosed, recovered, and deceased. The authors also assumed that the isolation and quarantine strategy was applied without glitch, which—again—is not fully realistic. In the medical world of epidemics, rules are difficult to implement by the book. Still, as the SARS experience in Vancouver and Toronto showed, they can come close to perfection, so this assumption of the model was not too far from what can be achieved in practice.

Under these circumstances, the authors computed that policies such as those used in Toronto would reduce the size of the SARS outbreak by a factor of 1,000. In other words, instead of the 300 infected individuals in the first month after the detection of the disease, Canada would have had to cope in the long run with at least 300,000 cases, but more likely three or four times more. These numbers speak for themselves of what could have happened without the intervention of the Canadian medical authority.

Vaccine Cocktails

In 1998, Michael Deem, then an assistant professor in bio-engineering at the University of California in Los Angeles, received a flu shot at his local Costco store. Before he left the site, the nurse advised him to repeat the procedure the next year because otherwise he would be more susceptible to infection than if he didn't get the shot at all. Though Deem knew what she was talking about, he had never thought about this problem in detail. The nurse's remark prompted a question in his mind.

Vaccines are standard measures against influenza. Small, inoffensive amounts of one or several strains are injected into the body, which reacts by building antibodies—the 'armies' needed to destroy a real invasion. The problem is that these antibodies also inhibit the formation of the antibodies needed against a new virus. So to be effective, vaccines have to replicate as close as possible next year's influenza strains. The experts who design the vaccination cocktail must therefore predict many months in advance the best composition of the shot. They need to forecast how the current viruses will mutate in the near future.

The classical procedure of the World Health Organization and the Centers for Disease Control and Prevention is to select the main existing strains, cultivate them in chicken eggs for four months, and then test them on ferrets. The reaction of the animals to the vaccine determines whether the combination will be used on humans. But there is no guarantee that humans will react exactly as the ferrets, so the vaccines are not always effective. Indeed, between 30 and 40 per cent of the elderly people who get vaccinated may still develop the disease. The cocktail is somewhat more effective in the young. Still, it is much less successful than most people assume.

Although the common flu is not much publicized, it leads to a lot of deaths among young children and seniors, killing between a quarter and half a million individuals every year worldwide. In the United States alone, more than 40,000 residents die of influenza, with a cost to the economy of ten billion dollars. The question Deem asked was aimed at fixing this problem. He wanted to know whether he could find a way to predict the best composition of the cocktail.

In the following years, Deem secured tenure and promotion, and then accepted a chaired position as John W. Cox Professor at Rice University in Houston, Texas. He conducted research on various topics related to the immune system, from HIV and dengue fever to influenza. And he also answered the question he had asked in 1998. At a meeting held in Los Angeles in March 2005, Deem reported about a method that appears to predict the best cocktail of flu strains. He didn't need the ferrets for this purpose. His approach used statistical mechanics.

Deem noted that the immune system is a biological example of a random, evolving structure, which has three key features. First, everybody has a finite number of antibodies out of a large number of possibilities. Second, the immune system must defeat a disease in a finite time. Third, there is a correlation between the immune system and the infection, because diseases evolve in response to the immune system. Such dynamical properties are encountered in the problems of statistical mechanics, which Deem mastered well due to his previous training in physics and the underlying mathematical background. His next step was to come up with an approach that would allow him to apply his expertise.

Using probability theory and statistics, Deem and his colleagues Jun Sun and David Earl managed to construct and analyse a model that provided them with, so far, the best possible combination of viruses for the yearly influenza vaccine. Their calculations showed that the efficiency rates they could obtain if their method was implemented might limit the failure of the vaccine to between 8 and 20 per cent of the cases, much better than the current 30 to 40 per cent rate.

Deem's research showed that the strain called Wyoming, of the 2004 vaccine, should be replaced with the related Kumamoto strain. Instead the 2005 vaccine replaced Wyoming with an emerging strain named California. According to Deem's model, this was a good choice too. These findings provide another example of how mathematics made a prediction that limits the spread of disease.

The Next Pandemic

The medical world has no doubts that a pandemic will grip the planet sometime in the future. Nobody, however, can predict when. The genetic mutation of viruses and the immune system are not well enough understood to allow accurate forecasts. But as presented in this chapter, we are not clueless about what to expect and how to confront such an event. We know some of the viruses that have to be watched, and we have learned what measures to take to contain the disease.

Though far from telling us everything we would like to know about a future pandemic, science is providing more information on what we need to do than what we can actually achieve through social and economic means. The poor countries of the world have little or no chance of defending themselves against a global epidemic, and even the rich nations keep insufficient vaccines to satisfy the needs of their citizens.

Many of the prevention measures are too costly to be implemented even in the western world. Vaccines, for instance, are expensive and expire quickly. Still, many of the medical resources are often wasted or not efficiently used. The United States of America, the richest country in the world today, has fifty million people without medical coverage. It is probably a good guess that the next pandemic will affect these people more than those who are insured.

Even the countries with a universal medical system are not fully prepared for the challenge. In Canada, for instance, federal and provincial politicians have not agreed who is responsible for certain costs, what segment of the population to vaccinate, and how to react in the small details if a pandemic hits the country. Each level of government is trying to shift responsibility on the other. Unfortunately such political skirmishes exist everywhere, and they stand in the way of dealing with epidemics.

So before asking more from science, which does its best to alleviate the impact of a global infectious disease, we should try to match our social, economic, and political efforts to what science has achieved so far. Indeed, it is not enough to know what we have to do; we must also find the means to do it. Otherwise it would be like predicting a collision with a comet, but having no plan to avoid it.

Of course, this doesn't mean that scientists can shift responsibility and enjoy their local victories. There is still much work to do, a fact of which they are fully aware. What they are not always sure about is whether they go in the right direction. Like the chief bankers who brought the world to the brink of a global financial collapse in 2008, scientists are not immune from making wrong decisions. So before ending our accounts of predicting megadisasters, let us examine the role mathematical models play in science and the ethical issues all experts face in their everyday research.

9. MODELS AND PREDICTION: HOW FAR CAN WE GO?

What distinguishes a mathematical model from, say, a poem, a song, a portrait or any other kind of 'model,' is that the mathematical model is an image or picture of reality painted with logical symbols instead of with words, sounds or watercolors.

John Casti

I began to like mathematics when I was fourteen years old. Before that, I did well in my classes, but without feeling much enthusiasm for it. My passion for the subject sprouted when I discovered geometry.

I had been taught in school that every mathematical statement needs a proof. Our teacher emphasized the difference between *hypothesis*, or the assumptions of the problem, and *conclusion*, or what we had to prove. Geometry suited this principle because it was easy to write down the hypothesis and the conclusion. Finding the proof was the hard bit.

I soon filled a notebook with solutions of geometry problems, and my mathematical skills improved. Often I found myself thinking of some difficult problem at night before falling asleep or in the morning while not fully awake. Nothing could beat the satisfaction of seeing a solution. That's when I decided to become a mathematician.

Though my initial interest was in purely mathematical questions, I soon began to look at applications. What appealed to me was the power of mathematics to investigate fields that have apparently nothing to do with it. Differential equations, the branch of mathematics in which I specialized, often finds such unexpected applications.

Through differential equations, I got acquainted with mathematical modelling. But soon I understood that finding a good model is an enterprise of which the most challenging part is staying in touch with reality.

The Emperor Has No Clothes

In 1968, the French mathematician René Thom published a paper in which he laid the foundations of catastrophe theory. Four years later he detailed the subject in a book. Thom had been well known in the mathematical community for a decade, since winning the Fields Medal—the most important award in mathematics until 2002 when Norway established the one million dollar Abel Prize. The scientific success of his book was therefore expected. Its appeal beyond the academic world, however, came as a surprise.

Thom studied how certain functions vary when the parameters they contain are perturbed. This variation is usually continuous, but the function leaps now and then. Thom called these jumps 'catastrophes', a choice of name, which together with the potential applications of the theory, attracted media attention.

Another famous mathematician joined René Thom in developing the applications: Christopher Zeeman—a charismatic professor from the University of Warwick in England. In a conversation we had in my office, where a map of the tectonic plates adorns one of the walls, he compared the Earth's crust with the skin that floats on the surface of a milk pot, a metaphor I used in Chapter 2.

Zeeman applied catastrophe theory to a large number of disciplines, including physics, sociology, economics, biology, linguistics, and psychology, and the media started to publicize his works. Newspaper and magazine articles claimed that we would soon be able to foretell market crashes, earthquakes, heart attacks, the downfall of dictatorships, and many other extreme events. Catastrophe theory was called the most important scientific achievement of the millennium.

But the media failed to mention that the theory was merely descriptive. All it did was to point out under what circumstances jumps occur, and none of its conclusions went beyond common sense. Though he implied that there was forecasting potential, Zeeman never predicted the place and magnitude of an earthquake or the fall of the Berlin Wall. He only pointed out that such things could happen when certain conditions are satisfied. Often those circumstances had been well known in the applied fields.

A much discussed example was the use of the 'cusp catastrophe' to analyse the stock market. The model portrays a cusped surface resulting from the graph of a certain function, and shows that crashes and slow recoveries follow after unreasonable market peaks. The model cannot predict the date of any crash; it only explains what happens when the market exceeds the expectations.

Of course, this statement is not new. Everybody knows that bull markets may be followed by crashes and slow recoveries.

So in this case, as in all the examples it produced, catastrophe theory only gave geometrical interpretations of well-known facts. Lacking any spectacular breakthrough, this branch of mathematics had not come even close to the realm of the top scientific achievements of the millennium. But until 1976, the media kept praising it as such.

The mood, however, changed quickly after several mathematicians raised concerns about the validity of the models. Héctor Sussmann of Rutgers University, John Guckenheimer from the University of California in Santa Cruz, and Stephen Smale of Berkeley pointed out that although the mathematics behind the theory is fine in general, the models used had little to do with the reality they pretended to describe. Bear markets, for instance, are not always followed by crashes, as the 'cusp catastrophe' model suggested, and similar weaknesses could be found in other applications. The public blow came from the journal *Science*, which in April 1977 published an article entitled *Catastrophe Theory: The Emperor Has No Clothes*.

Thom answered the criticism without apology. After all, he and Zeeman had never pretended that the proposed models had capability of prediction. 'Catastrophe theory is not a "scientific theory," ' Thom wrote. 'It is a language and, as with ordinary language, every author will use it to his own taste and with his own "style." ' What he meant was that, as in its stock-market model, catastrophe theory could explain some facts in geometric terms, thus giving an intuitive image of the underlying phenomena. It did not exhaust all possibilities and had no pretence for predictions.

Was this the end of the field? No. Catastrophe theory is still researched today, and new mathematical results are published in specialized journals. But its aura vanished in 1977. After years of unmerited praise, followed by a period of undeserved trashing, this branch of mathematics found its rightful place among the other scientific theories. As for its applications, the domain is viewed now exactly as Thom characterized it: a language that can describe facts but has no power of prediction. The use of 'mathematical model' in this context is therefore inappropriate because mathematical models are independent of language and should allow forecasting.

Models and Reality

The example of catastrophe theory raises natural questions: How good are the models of science? Can we rely on them for deterministic predictions? The answers vary with each modelled phenomenon. Some, like the motion of celestial bodies, are highly reliable for long intervals of time. Others, as the weather, allow only short-term forecasts. Yet others, such as earthquakes, are still elusive with no immediate perspective of improvement.

Regularity in the motion of celestial bodies has been recognized and used since ancient times to predict eclipses and compute planetary orbits. But Edmund Halley's success to foretell the return of a comet, which occurred in 1757 as he had predicted, showed the value of Newton's gravitational model. The discovery of Uranus and Neptune through hand-made calculations confirmed its robustness, thus making many eighteenth- and nineteenth-century mathematicians suspect that the solar system was stable. The French mathematicians

Laplace, Lagrange, and Poisson even showed that within certain approximations the motions of the planets were stable as hoped.

In 1878, however, a Romanian mathematician named Spiru Haret proved in his doctoral thesis defended at the Sorbonne that, in a better approximation than the one of his predecessors, the answer was inconclusive. The occurrence of certain terms in his more precise solutions stopped him from confirming that the system was stable. Haret was the first who raised some doubt about the reliability of the model, because in the absence of stability the problem of predicting the distant future becomes notoriously difficult.

Just as Haret reached this conclusion, a young Henri Poincaré had started working on his dissertation at the same Sorbonne. There were no signs yet that he would become one of the best mathematicians ever. But he soon proved himself. In 1886 he was already so famous that he received an invitation to take part in a contest whose main goal was to produce a solution of the N-body problem—the mathematical framework that generalizes the solar system. The jury had recognized the importance and difficulty of this problem and wanted to stimulate Poincaré to solve it. He was among the few mathematicians who could make a breakthrough.

King Oscar II of Sweden and Norway, a great supporter of culture, sponsored this event, whose prestige equated today's Nobel Prize. The winner was to be announced on 21 January 1889, which coincided with the king's sixtieth birthday. Poincaré accepted the challenge and started working on the problem. He couldn't solve it completely in spite of more than two years of work, but obtained such a breadth of results and shed so much light on it that the jury awarded him the prize with enthusiasm. No other competitor had come even close to his level.

Among other things, Poincaré claimed to have proved the stability of the solutions describing the problem, and consequently of the solar system. But months after the competition had closed, he found a mistake in his proof. Moreover, he understood that he could not fix it. He had begun to cast serious doubts whether the solutions were stable.

Poincaré was a man of great character, and he immediately telegraphed to Stockholm to stop the printing of *Acta Mathematica*'s volume 13, the journal that would publish his paper. He offered to reimburse the costs of printing, which were substantially more than the prize money he had received. Unfortunately some copies had already been sent to subscribers, and a scandal appeared imminent.

But Gösta Mittag-Leffler, the editor-in-chief of the journal and the scientific advisor to King Oscar II, managed to calm things down, while Poincaré engaged in correcting the result. In the months that followed, under tremendous pressure, the French mathematician discovered the chaos phenomenon. Though he did not prove that the N-body problem is chaotic, he couldn't show that it wasn't. A complete argument still eludes us today, though we have ample evidence of the system's chaotic behaviour.

Luckily, chaos acts slowly in the solar system. It takes the planets hundreds of millions of years to drift away from the current orbits. This is the reason why we can send missions in space without any worry of miscalculating their trajectories. We should, therefore, have no problem with computing the orbits of the asteroids or comets that might hit Earth. The reason why we can't do it immediately after such an object is discovered lies with insufficient data. Only after observing it long enough can we determine its orbital elements and mass. But the model is sound for our immediate purposes, and chaos doesn't affect it in the short term.

Unfortunately we cannot say the same about the weather. Although the models we use for the atmosphere have improved in recent years, they still don't capture all the essential details of the physical reality. The formation and dissipation of clouds, for instance, eludes our understanding. Clouds still become part of most models but in parametric terms, which means that they are introduced through empirical equations. Such models can be tuned to fit the past, but are hard to make to predict the future.

Apart from this difficulty, all the weather models are chaotic too, with rates of instability that can be felt within days. This is the reason why a week-long weather forecast only rarely succeeds, and when it does it is rather by luck. Unlike celestial bodies, the weather is not prone to precise and robust dynamical models. The shorter the time interval, the more reliable the forecast. For hurricanes, the experts can give good warnings two to three days in advance, just enough time to evacuate small and medium communities. Big cities, however, are at higher risk.

Some recent research shows that chaos is less responsible for the errors of the current meteorological forecasts than are the models themselves. This is the point David Orrell makes in his recent book, *Apollo's Arrow—The Science of Prediction and the Future of Everything*. His conclusions are based on the results of his doctoral dissertation defended in Oxford, England. Nevertheless, chaos is still a factor of computational error for many classes of differential equations, as mathematicians have proved in a large body of work performed during the past decades.

Finally, the most elusive predictions are those based on vague models, as it happens in seismology. Without knowing the exact position of tectonic plates and the way they move, we are unable to come up with precise models. So far, the best we can do is to enforce building codes in unstable seismic zones. Pandemic models are also not good enough to foretell when a virus might strike a large segment of the population, but at least the experts can warn us about where to look and what measures to take to stop or contain the spread of a disease.

Apparently mathematical models should approach reality as closely as possible. But this condition cannot be applied to the study of complex systems, which have so many parameters that the equations describing them become too complicated to be of any use. For example, a model of the stock market that captures the reaction of every investor would be impossible to build or handle. Therefore approaches that look at patterns and tendencies, but ignore other details, are among the ones with better chances to succeed. The models of Didier Sornette, discussed in Chapter 7, are of this type, and they have shown some degree of success in predicting the evolution of the market.

Similarly, the atmosphere is a complex system, so climate models cannot come very close to reality. But unlike weather forecasting, which is a difficult enterprise due to the sensitivity of the equations that describe the phenomenon, the global average temperature of the planet can be obtained. Although the estimates are not highly accurate, all the existing models show the same tendency of temperature increase for the rest of this century. Were the models wrong, their predictions would likely disagree with each other.

These remarks show that in some cases models must be close to reality, while in others they cannot or don't have to be. The nature of the studied phenomenon is what must be considered when deciding what kind of mathematical model to adopt.

The Mirage of the Crystal Ball

Scientists are sometimes pressed to issue forecasts in exchange for grants, which are essential for their research. Robert Geller, who was mentioned in Chapters 1 and 2, complained about this problem in the field of seismology. He is not alone. Orrin Pilkey, an emeritus geology professor at Duke University, and his daughter, Linda Pilkey-Jarvis of Washington's Department of Ecology, raised the issue of model abuse in environmental studies. Their book, *Useless Arithmetic—Why Environmental Scientists Can't Predict the Future*, published by Columbia University Press in 2007, analyses some serious ethical issues.

For the past twenty-five years, the authors had monitored beach nourishment projects in the United States. They noticed that to obtain federal grants for their projects, the experts had to predict how long the sand would last on the beach. The trouble was that those predictions were wrong most of the time, always erring towards whatever arguments would bring in funding. When asked to explain, the experts blamed unexpected storms for the errors. But there is nothing unusual about storms. They occur often, and any prediction must take them into account.

'Do such things happen in other branches of environmental studies?' the authors asked. They spent a few more years looking into this question and found that such practices are widespread. Among the few sound research directions were hurricane tracking and climate modelling. Most of the other fields seemed to have gone astray. The standard was so low that the experts didn't even check after the fact whether their predictions had been correct. As in the case of catastrophe theory, the Pilkeys didn't blame the mathematics for this outcome, only the sloppy assumptions behind the models.

Among their several case studies was the North Atlantic Grand Banks fishery, known as the main world source of cod in the past five centuries. Unfortunately, its industry collapsed in 1992 as a result of overfishing. More than 40,000 people lost their jobs. This disaster resulted from decisions based on faulty mathematical models, which had highly overestimated cod resources. But since the word mathematics was used to justify those decisions, all the dissenting voices were silenced. Such unethical practices do not only destroy the environment, but harm science as well.

Another case the authors considered was the choice of the Yucca Mountain in Nevada as a permanent site of America's nuclear waste. A mixture of several models guaranteed that no radiation would leak from there in the next million years to endanger local communities. The authors found this claim absurd to the degree of irresponsibility. How can anybody guarantee what will happen in the next million years, in which earthquakes, continental shifts and ruptures, volcanic eruptions, change of landscape, ice ages, or periods of very hot climate could release nuclear radiation and endanger all the life in the area?

The two researchers found such models worse than useless because they provide 'a false sense of security and an unwarranted confidence in our scientific expertise'. How can funding be awarded to such research? The authors blame politics for that, since political bodies are the ones that provide the funds even if they give a scientific image to the decision by peer reviewing the scientific proposals.

Deeper behind such practices is our naive hope that we can see the future in crystal balls and therefore improve our lives. How many of us have never read the horoscope in the daily newspaper or have totally dismissed the connection between the zodiac in which we were born and the traits of our character? No wonder that the public doesn't complain about money awarded for dubious prediction models. The book by Pilkey and Pilkey-Jarvis is a wake-up call aimed at pointing the finger at some of those who use pseudo-science for questionable goals.

Within the limits of our scope, it is reassuring to know that other people appreciate the work done on hurricane tracking and climate forecasts. But at the same time *Useless Arithmetic* emphasizes that there is good and bad science, and that honest scientists must blow the whistle whenever something dubious happens in their field.

Unfortunately this kind of action is rare. Science lives in enclaves, and it is often difficult to judge the norms of your own field. Even those who understand what is happening are often reluctant to stir the pot because whistleblowers often hurt themselves rather than improve the situation. Moreover, as Aristotle put it, man is a political animal, and scientists are people too. Bart Kosko, a professor at the University of Southern California, spelled this idea in more detail in his book on fuzzy logic:

Career science, like career politics, depends as much on career maneuvering, posturing, and politics as it depends on research and the pursuit of truth... Politics lies behind literature citations and omissions, academic promotions, government appointments, contract and grant awards... and, most of all, where the political currents focus into a laser-like beam, in the peer-review process of technical journal articles—when the office door closes and the lone anonymous scientist reads and ranks his competitors' work.

Kosko is right up to a point. What he doesn't say is that scientists have a conscience too, and they rarely ignore brilliant results on purpose. After all, what experts fear most is losing their credibility. Diamonds shine even in the mud, and if you don't see a diamond when it's right in front of you, there is a good chance that others will recognize it sooner or later. This is the reason why science makes progress. If only politics and envy reigned among us, we would still be rubbing wood sticks to kindle a fire.

The moral of the tale is that some experts do abuse science for selfish purposes, but they are exceptions. Most scientists are striving to make justifiable claims. The same happens with those who issue forecasts. Predictions usually come with a warning about how approximate the results are. Unfortunately such details get lost through the filter of the media, and the public learns only the breaking news.

The successes mentioned in the past chapters would have been impossible without the scientists who try to understand the phenomena that surround us. When compared to the recent past, these men and women have made huge progress in their quest to predict megadisasters. Indeed, any vigilant person could recognize an incoming tsunami; hurricanes don't hit without warning anymore; volcanic eruptions have lost their mystery; we know that the climate is warming; we watch the evolution of viruses to avoid a pandemic; we discover and follow the celestial objects that might hit us and look for ways to avoid a collision.

Of course, we are still far away from predicting all mega-disasters, including earthquakes and stock-market crashes. But science makes progress, and its mathematical models are improving. Still, nobody can tell whether we will ever feel fully secure when confronted with the forces of nature. Certain, however, is that scores of people all over the globe are trying to make this world a safer place.

This book is dedicated to them.

ACKNOWLEDGMENTS

Many people helped me with this book. I would like to acknowledge my editor Latha Menon and copyeditor Jess Smith, as well as the dedicated staff at Oxford University Press: Fiona Vlemmiks, Emma Marchant, Sara-Louise Cain, Clare Hofmann, and Claire Thompson; my agent Samantha Haywood; my friends from Victoria: Elisabeth von Aderkas, Maggi Feehan, Margaret Gracie, Kitty Hoffman, and Paul Mohapel; all my reviewers, including Markus Brunnermeier, John Dvorak, Kerry Emanuel, Chris Garrett, Perry Kendall, John S. Lewis, Didier Sornette, Neil deGrasse Tyson, Harry Yeh, and the late Alwyn Scott; and, last but not least the people I share my life with: my partner Mariana Diacu, who enjoyed reading the first draft of this book as much as I enjoyed writing it, and our son, Răzvan, who helped me with my research and produced the index.

A Note About the Type

The text of this book is set in Adobe Garamond, a modern interpretation of the historical types of Claude Garamond and Robert Granjon, designed by Robert Slimbach in 1989. His type captures the beauty and elegance of the original classical Garamond typefaces. The display font, used for chapter headings, is a modern, edgy, distressed typeface called Down come. It was designed by Eduardo Recife in 2002 and is freely available from www.misprintedtype.com.

NOTES

p. vi 'You can only predict . . .' Eugène Ionesco, *The Rhinoceros*, Act 3.

p. vi 'Predicting the future is easy . . .' In [Lewis 1987], journalist Peter H. Lewis quotes Fritz R. S. Dressler, president of FRS Dressler Associates of Wallingford, PA, writer, and computer consultant, as saying while on a panel discussing the future of Apple computers.

p. vi 'We have redefined the task . . .' Stephen Hawking, *A Brief History of Time*, ch. 11.

Prologue: Glimpsing the Future

p. 1 'Predicting is very difficult . . .' is attributed to physicist Niels Bohr but also to baseball player and manager Yogi Berra, known for his provocative language.

1. Walls of Water: Tsunamis

p. 7 Dave Lowe's quote 'I got outside my hotel, . . .' is from his blog, which appears at <http://phukettsunami.blogspot.com/2005/12/survivor-dave-lowe.html>.

p. 8 The transcript of Petra Nemcova's interview on Larry King Live appears at <http://transcripts.cnn.com/TRANSCRIPTS/0505/13/lkl.01.html>. The story of Petra Nemcova and Simon Atlee is also told in [Nemcova 2005].

p. 11 The media reported largely about Richard Gross's results concerning the tilt of the Earth's axis. An Internet search reveals hundreds of sources; see, e.g., <http://www.ammas. com/topics/Current_Affairs/a99843.html>.

p. 11 The conclusion of Seth Stein and Emile Okal appeared in a press release from Northwestern University and was reported widely in the media; see, e.g., Michael Shirber's article, 'Tsunami Earthquake Three Times Larger Than First Thought', published in *Live Science* on 8 February 2005: <http://www. ammas.com/topics/Current_Affairs/a99843.html>.

p. 12 A good article that covers the mathematics of many achievements related to the solitary wave is [Darrigol 2003]. See also [Bullough 1988], [Craik 2004], and [Filippov 2000].

p. 12 Airy and Stokes mention the impossibility of solitary waves in [Airy 1841] and [Stokes 1880], respectively.

p. 13 'Accumulated round the prow of the vessel . . .' [Russell 1845: 312].

p. 13 'Followed it on horseback, and overtook it . . .' [Russell 1845: 312].

p. 17 'I have in mind . . .' [Darrigol 2003: 66].

p. 17 'To see and feel the waves' [Darrigol 2003: 66].

p. 17 'You ask if I have done . . .' [Darrigol 2003: 66].

p. 18 'Contrary to an opinion . . .' [Darrigol 2003: 66].

p. 19 'An infinitely fluid mass, initially at rest . . .' in *Procés-verbaux de l'academie des sciences*, 5 (1812–15), 262 (author's translation from French).

p. 20 Boussinesq's paper is [Boussinesq 1871].

p. 22 'The new type of long stationary waves' [Korteweg 1895].

p. 22 The work of Lord Rayleigh appeared in [Rayleigh 1876].

p. 26 The works of Fermi, Pasta, Ulam, Zabuski, and Kruskal are described in [Darrigol 2003]; the paper by Fermi, Pasta, and Ulam is [Fermi 1955].

p. 28 The work by Hills and Goda that computes the size of tsunamis triggered by meteoritic impacts is [Hills 1993].

p. 30 The event involving Tilly Smith has been widely reported in the media; see, e.g., <http://news.bbc.co.uk/2/hi/uk_news/4229392.stm>.

2. Land in Upheaval: Earthquakes

p. 34 'If a seismologist fails to predict...' appeared in the *Smile* section of *The Globe and Mail* on 25 January 2007, p. 2.

p. 37 The article quoting Garry Rogers is [Lavois 2007].

p. 37 'It looks like we dodged...' an editorial in *The Times Colonist* of 6 February 2007.

p. 39 More on the history of seismology can be found in [Davison 1978].

p. 43 'In a more determinate manner...' [Davison 1978: 68].

p. 43 'The sudden flexure and constraint...' [Davison 1978: 69].

p. 44 'Though seismology provides...' Introduction of [Stein 2003].

p. 45 'The professional duty of a scientist...' [Dyson 1988: 258].

p. 48 'There is no particular reason...' Interview with Robert Geller taken by Eric A. Bergman in 2001 for GeoNavi.

p. 46 The drilling of the 12.262-kilometre-long hole, done for the pure reason of seeing how deep we can go, took place in Russia's Kola Peninsula, near the border with Norway, between 1970 and 1994.

p. 48 'No less of a scientist...' [Geller 1999].

p. 49 'Sesimologists would like to be able to...' [Hough 2002: 2].

p. 49 'Since my first attachment to seismology...' [Geller 1997: M5] or [Richter 1977: 1].

p. 52 The case of Bailey Willis is discussed in [Visher 1949].

p. 52 'Is the reliable prediction...' is the title of Ian Main's article that opens the *Nature* debates, and was published in the issue of 25 February 1999.

p. 53 A discussion of the work by an American-Indian research team, which attempted to identify a precursor by looking at algae blooming in the coastal waters of British Columbia in June 2006 appears in [Hartley 2006].

p. 54 'So if we cannot predict...' *Nature*, Ian Main, 25 February 1999.

p. 54 'I am pessimistic about the near future...' *Nature*, Max Wyss, 25 February 1999.

p. 55 'No kind of prediction is more obviously mistaken...' *Nature*, Andrew Michael, 25 February 1999.

p. 55 'Refrain from using the argument...' *Nature*, Robert Geller, 11 March 1999.

p. 56 'It is not earthquakes themselves...' *Nature*, Ian Main, 8 April 1999.

p. 58 'A blend of confusion, empirical analysis...' [Wang, 2006: 794]. For further reading on this topic see [Mason 1968] and [Hough 2002], to appreciate the progress made in thirty-four years.

3. Chimneys of Hell: Volcanic Eruptions

p. 61 'The cloud was rising from a mountain...' letter from Pliny the Younger to Tacitus; see, e.g., [Mellor 2004: 532] or <http://www.volcanolive.com/pliny.html>.

p. 62 'During the past 400 years...' [De Boer 2002: 2].

p. 64 A good source about the Krakatoa event is [Winchester 2003].

p. 64 The paper of Chris Garret is [Garrett 1969].

p. 65 The description of the eruption of Mount Pelée appears in [De Boer 2002: 186–208].

p. 65 'The bottom of the pit all red...' <http://www.answers.com/topic/havivra-da-ifrile>.

p. 66 A history of volcanology appears in [Sigurdsson 1999].

p. 70 Raspe's story is told in [Sigurdsson 1999: 140–7].

p. 71 'An accumulation of ashes thrown...' is from Raspe's *The Surprising Adventures of Baron Münchausen*, [Raspe 1895], which can also be found at <http://www.fullbooks.com/The-Surprising-Adventures-of-Baron-Munchausen2.html>.

p. 71 'The appearances of coal and cinder...' same source as previous quote.

p. 72 'Mount Vesuvius was another...' same source as previous quote.

p. 74 The results of the research done at Soufriére Hills in Montserrat appear in [Sparks 2001], and can be found at <http://www.firstscience.com/home/articles/earth/dynamic-pulsating-eruptions_1266.html>.

p. 75 For the probabilistic results related to eruptions at Soufriére Hills in Montserrat see [Connor 2003].

p. 78 Plenty of information on the lahar catastrophe at Nevado del Ruiz in Colombia can be found on the websites of the US Geological Survey.

p. 84 The website of the US Geological Survey is a good source for the events that led to the birth of Paricutin in Mexico.

p. 86 Most information about Katja and Maurice Krafft is taken from their obituary, [Keller 1992].

p. 87 'Thank you Maurice and Katja...' [Keller 1992: 614].

4. Giant Whirlwinds: Hurricanes, Cyclones, and Typhoons

p. 89 'I am more afraid...' The US President William McKinley made this remark at the turn of the twentieth century, after Willis Moore of the US Weather Bureau stated that hurricanes had sunk more ships than all naval wars, [Elsner 1999: 49].

p. 89 An excellent source for this chapter, as well as a highly recommended further reading, is [Emanuel 2005].

p. 94 'No mere photograph can do justice...' [Emanuel 2005: 195].

p. 96 'It is to heat we should attribute...' is Kerry Emanuel's translation of Carnot's quote, which appeared in *Reflexions sur la Puissance Motrice du Feu* (Paris, 1824), see [Emanuel 2005: 54].

p. 106 'My dad, a pilot, was away on a trip . . .' appears in <http://www.memoryarchive.org/en/Hurricane_Elena,_1985,_by_Ashley_Wroten>.

p. 107 The story of the first airplane flight into a hurricane is told in the Weather Almanac of July 2003; see <http://www.islandnet.com/~see/weather/almanac/arc2003/almo3jul.htm>.

p. 117 'As the Atlantic hurricane season gets underway . . .' is signed by Kerry Emanuel, Richard Anthes, Judith Curry, James Elsner, Greg Holland, Phil Klotzbach, Tom Knutson, Chris Landsea, Max Mayfield, and Peter Webster, and appears at <http://wind.mit.edu/~emanuel/Hurricane_threat.htm>.

5. Mutant Seasons: Rapid Climate Change

p. 119 'Our response to the threat of global warming . . .' [Weart 2003: viii]. The IPCC reports can be found at <http://www.ipcc.ch/>.

p. 120 'Because concern about climate change . . .' [Flannery 2005: 4].

p. 121 The story of the discovery of global warming is told in [Weart 2003].

p. 123 I provided the correct, 26,000-year period, for the 'top-like wobbling' of the Earth's motion called precession. Milankovich gives a shorter estimate, namely 21,000 years.

p. 123 Charles Brooks's book is [Brooks 1926].

p. 125 Velikovsky's two cited books are [Velikovsky 1950] and [Velikovsky 1955]. For a critical analysis of Velikovsky's ideas see [Diacu 2005: ch. 1].

p. 125 The story of the research behind the theory of the disappearance of the mammoths because of an exploding star is told in [Boswell 2005: A12].

p. 133 About research done on brown clouds see [Ramanathan 2006].

p. 137 An excellent book that analyses the connection between hurricane activity and climate change is [Mooney 2007].

p. 139 'Perhaps the best advice chaos theory...' [Lorenz 1991: 450].

p. 143 'It is about time that the Gulf-Stream-European climate myth...' [Seager 2006].

p. 146 'Over the past few years...' [Wente 2007].

p. 147 'Nobody can tell you...' [Wente 2007].

p. 147 'Upper bound for the sea-level rise...' [Cormier 2007].

6. Earth in Collision: Cosmic Impacts

p. 148 'Old men and comets...' Jonathan Swift, *Thoughts on Various Subjects*, 1727.

p. 149 The size of the largest fragment of the Schoemaker–Levy comet has been estimated from several hundred metres to three kilometres. Most researchers think the smaller estimates are more realistic than the larger ones.

p. 152 The history of the discovery of asteroids is well known. A brief presentation appears in [Verma 2005].

p. 153 'I have announced this star as a comet...' *Corrispondenza Astronomica fra Giuseppe Piazzi e Barnaba Oriani*, eds. G. Cacciatore and G. V. Schiaparelli (Milan: University of Hoepli, 1874), 204.

p. 155 The theory suggesting that the asteroid responsible for the death of the dinosaurs resulted from a collision between two larger asteroids appeared in [Bottke 2007]. The initial collision took place about 160 million years ago, some 100 million years before the impact with Earth.

p. 172 'Comets and asteroids remind me of Shiva...' [Gehrels 1996]. For further reading, see [Steel 1995], [Lewis 1996], [Verschuur 1996], and [Cox 1996].

7. Economic Breakdown: Financial Crashes

p. 179 'I think that the technician...' [Poundstone 2005: 308].

p. 179 'The key data is in my view...' [Poundstone 2005: 308].

p. 179 'If we try it and don't like it...' [Poundstone 2005: 308].

p. 180 The information about Robert Shiller is from [Campbell 2006].

p. 181 'But how do we know when irrational exuberance...' Alan Greenspan's discourse appears at <http://www.federalreserve.gov/boardDocs/speeches/1996/19961205.htm>.

p. 181 The book by Robert Shiller is [Shiller 2000]. Shiller's graph (shown in Figure 7.1) used the Standard and Poor's Composite Stock Price Index and the corresponding earnings, both adjusted to inflation.

p. 184 'There are crimes far worse...' [Hoover 1952: vol. 3, 14].

p. 187 'There is nothing now to be forseen...' [Sutliff 1925: 1].

p. 187 'A new age is taking form...' [Moody 1928: 260].

p. 187 'The rich man's chauffeur...' [Allen 1931: 315].

p. 188 'Stock prices have reached what looks...' [Shiller 2000: 106]. 'Most of the money formerly spent in the saloon...' [Hazelwood 1929: 31].

p. 191 'Very difficult to definitively identify a bubble...' is a remark made by Alan Greenspan, Chairman of the Federal Reserve Board, on 30 August 2002 at a symposium sponsored by the Federal Reserve Bank of Kansas City in Jackson Hole, Wyoming. His speech is posted on <http://www.federalreserve.gov/boarddocs/speeches/2002/20020830/>.

p. 192 The theory of Didier Sornette is presented in [Sornette 2003].

8. Tiny Killers: Pandemics

p. 200 'When we think of the major threats...' is a thought Barak Obama expressed during his campaign of 2007 for the

Democratic Party's nomination for the American Presidency; see <http://www.huffingtonpost.com/bruce-kluger-and-david-slavin/behold-the-new-barack_b_72521.html>.

p. 204 The age-mortality curve for the 1918 flu pandemic, which killed many adults aged 20 to 40, looked like a W, unlike the usual influenza curve, which rather resembles a V.

p. 208 The works of Ferguson mentioned here are [Ferguson 2003] and [Ferguson 2006].

p. 211 'Escaping the pandemic entirely...' [Morse 2007: 7314].

p. 213 The work on travelling one-dollar bills appears in [Brockmann 2006].
The work of Larson is described in [Larson 2007].

p. 215 'We allow for socially active people...' [Larson 2007: 400].

p. 216 'We provide for highly susceptible people...' [Larson 2007: 400].

p. 219 The work of Deem appears in [Sun 2006].

p. 220 A detailed description of how vaccine viruses are selected appears at <http://www.cdc.gov/flu/professionals/vaccination/ virusqa.htm>.

9. Models and Prediction: How Far Can We Go?

p. 225 'What distinguishes a mathematical model...' [Casti 1992].
p. 226 The book by Thom is [Thom 1972].
p. 227 Zeeman's popular-science article on catastrophe theory is [Zeeman 1976].
p. 228 'Catastrophe theory is not a "scientific theory"...' [Kolata 1977: 351].
p. 228 A classic text on catastrophe theory is [Poston 1978].
p. 230 The story of Poincaré's discovery of the chaos phenomenon appears in [Diacu 1996: ch. 1].

p. 232 The book by Orell is [Orell 2007].

p. 234 The criticism of environmental models appears in [Pilkey 2007].

p. 237 'Career science, like career politics,...' [Kosko 1993: 41–2]. Readers interested in a more mathematical treatment of natural catastrophes can consult [Woo 1999].

SELECTED BIBLIOGRAPHY

[Airy 1841] Airy, G. B. Tides and Waves, *Encyclopedia Metropolitana* 3 (1841).

[Allen 1931] Allen, Frederick Lewis. *Only Yesterday*, New York: Harper and Brothers, 1931.

[Boswell 2005] Boswell, Randy. Exploding Star Wiped Out Mammoths 13,000 Years Ago, *National Post* (3 November 2005).

[Bottke 2007] Bottke, William, F., Vokrouhlicky, David, and Nesvorny, David. An Asteroid Breakup 160 Myr Ago as the Probable Source of the K/T Impactor, *Nature* 449 (6 September 2007), 48–53.

[Boussinesq 1871] Boussinesq, Joseph. Théorie de l'intumescence liquide appellée 'onde solitaire ou de translation' se propageant dans un canal rectangulaire, *Comptes-rendus hebdomadaires des séances*, Académie des Sciences, Paris 72 (1871), 755–9.

[Brockmann 2006] Brockmann, D., Hufnagel, L., and Geisel, T. The Scaling Laws of Human Travel, *Nature* 439 (2006), 462–5.

[Brooks 1926] Brooks, C. E. P. *Climate Through The Ages*, New York: McGraw-Hill, 1926.

[Bullough 1988] Bullough, Robin. The wave 'par excellence', the solitary, progressive great wave of equilibrium of the fluid—an early history of the solitary wave. In *Solitons*, ed. M. Lakshmanan, New York: Springer Verlag, 1988.

[Campbell 2006] Campbell, John Y. *An Interview with Robert Shiller*. Cowles Foundation Paper No. 1148, Cowles Foundation for Research in Economics, Yale University, 2006.

[Casti 1992] Casti, John. *Reality Rules: I, The Fundamentals*. New York: John Wiley & Sons, 1992.

[Connor 2003] Connor, C. B. *et al*. Exploring links between physical and probabilistic models of volcanic eruptions: The Soufrière Hills Volcano, Montserrat, *Geophysical Research Letters* 30/13 (2003), 1701.

[Cormier 2007] Cormier, Zoe. Will Oceans Surge 59 Centimetres This Century—or 25 Metres? *The Globe and Mail* (Saturday, 25 August 2007), p. F9.

[Cox 1996] Cox, Donald W. and Chestek, James H. *Doomsday Asteroid: Can We Survive?* Amherst and New York: Prometheus Books, 1996.

[Craik 2004] Craik, A. D. D. The Origins of Water Wave Theory, *Annual Review of Fluid Mechanics* 36 (2004), 1–28.

[Darrigol 2003] Darrigol, Oliver. The Spirited Horse, the Engineer, and the Mathematician: Water Waves in the Nineteenth-Century Hydrodynamics, *Archive for the History of Exact Sciences* 58 (1983), 21–95.

[Davison 1978] Davison, Charles. *The Founders of Seismology*, New York: Arno Press, 1978.

[De Boer 2002] De Boer, Jelle Zeilinga and Sanders, Donald Theodore. *Volcanoes in Human History: The Far-Reaching Effects of Major Eruptions*, Princeton, NJ: Princeton University Press, 2002.

[Diacu 1996] Diacu, Florin and Holmes, Philip. *Celestial Encounters: The Origins of Chaos and Stability*, Princeton, NJ: Princeton University Press, 1996.

[Diacu 2005] Diacu, Florin. *The Lost Millennium: History's Timetables Under Siege*, Canada: Knopf, 2005.

[Dyson 1988] Dyson, Freeman. *Infinite in All Directions*, London: Penguin, 1988.

[Elsner 1999] Elsner, James B. and Kara, A. Birol. *Hurricanes of the North Atlantic*, New York: Oxford University Press, 1999.

[Emanuel 2005] Emanuel, Kerry. *Divine Wind: The History and Science of Hurricanes*, New York: Oxford University Press, 2005.

[Emmerson 1977] Emmerson G. S. *John Scott Russell: A Great Victorian Engineer and Naval Architect*, London: Murray, 1977.

[Ferguson 2003] Ferguson, Neil M., Galvani, Alison P., and Bush, Robin M. Ecological and Immunological Determinants of Influenza Evolution, *Nature* 422 (2003), 428–33.

[Ferguson 2006] Ferguson, Neil M. *et al.* Strategies for Mitigating an Influenza Pandemic, *Nature* 442 (2006), 448–52.

[Fermi 1955] Fermi, E., Pasta, J., and Ulam, S. Studies of Non-Linear Problems, Document LA-1940 (May 1955), in *Collected Papers of Enrico Fermi*, vol. II, University of Chicago Press, 1965, 978–88.

[Filippov 2000] Filippov, A. T. *The Versatile Soliton*, Boston: Birkhäuser, 2000.

[Flannery 2005] Flannery, Tim. *The Weather Makers*, Toronto: Harper Collins, 2005.

[Galbraith 1979] Galbraith, John Kenneth. *The Great Crash 1929*, Boston: Houghton Mifflin, 1979.

[Garrett 1969] Garrett, C. J. R. A Theory of the Krakatoa tide gauge disturbances, *Tellus* 22/1 (1970), 43–52.

[Gehrels 1996] Gehrels, Tom. Collisions with comets and asteroids, *Scientific American* (March 1996), 55–7.

[Geller 1997] Geller, Robert. Earthquake Prediction Is Impossible, *Los Angeles Times* (2 February 1997), M5.

[Geller 1999] Geller, Robert. Earthquake prediction: is this debate necessary? *Nature* 25 (February 1999).

[Hankins 1970] Hankins, T. L. *Jean d'Alembert—Science and the Englightenment*. Oxford: Clarendon Press, 1970.

[Hartley 2006] Hartley, Matt. Canadian Scientists Debunk Algae-Quake Link, *Times Colonist* (7 July 2006).

[Hazelwood 1929] Hazelwood, Craig B. Buying Power Termed Basis for Prosperity, *New York Herald Tribune* (2 January 1929).

[Hills 1993] Hills, J. G. and Goda, M. P. The Fragmentation of Small Asteroids in the Atmosphere, *Astronomical Journal* 105 (1993), 1114–44.

[Hoover 1952] Hoover, Herbert. *The Memoirs of Herbert Hoover*, London: Hollis and Carter, 1953.

[Hough 2002] Hough, Susan Elisabeth. *Earthshaking Science: What We Know (and Don't Know) about Earthquakes*, Princeton, NJ: Princeton University Press, 2002.

[Keller 1992] Keller, Jörg. Memorial for Katja and Maurice Krafft, *Bulletin of Volcanology* 54 (1992), 613–14.

[Kolata 1977] Kolata, Gina Bari. Catastrophe Theory: The Emperor Has No Clothes, *Science* 196 (1977), 287 and 350–1.

[Korteweg 1895] Korteweg, Diederik and de Vries, Gustav. On the Charge of Form of Long Waves Advancing in a Rectangular Canal, and on a New Type of Long Stationary Waves, *Philosophical Magazine* 39 (1895), 422–43.

[Kosko 1993] Kosko, Bart. *Fuzzy Thinking: The New Science of Fuzzy Logic*, New York: Hyperion, 1993.

[Larson 2007] Larson, Richard C. Simple Models of Influenza Progression within a Heterogenous Population, *Operations Research* 55 (2007), 399–412.

[Lavois 2007] Lavois, Judith. Scientists warn of higher quake risk, *Times Colonist* (2 February 2007), 1.

[Lewis 1987] Lewis, Peter H. The Executive Computer: Apple Invades I.B.M.'s Domain, *New York Times* (30 August 1987).

[Lewis 1996] Lewis, John S. *Rain of Iron and Ice*. New York: Addison-Wesley, 1996.

[Lorenz 1991] Lorenz, Edward N. Chaos, Spontaneous Climatic Variations and Detection of the Greenhouse Effect, in *Greenhouse-Gas-Induced Climatic Change: A Critical Appraisal of Simulations and Observations*, ed. M. E. Schlesinger, Amsterdam, New York, Oxford, Tokyo: Elsevier Science, 1991, 445–53.

[Mason 1968] Mason, R. G. Can Earthquakes Be Predicted? Inaugural Lecture, Imperial College of Science and Technology, University of London, 1968.

[Mellor 2004] Mellor, Ronald. *The Historians of Ancient Rome: An Anthology of the Major Writings*, New York: Routledge, 2004.

[Moody 1928] Moody, John. The New Era in Wall Street, *Atlantic Monthly* (August 1928).

[Mooney 2007] Mooney, Chris. *Storm World*, New York: Harcourt, 2007.

[Morse 2007] Morse, Stephen S. Pandemics Influenza: Studying the Lessons of History, *Proceedings of the National Academy of Sciences* 104 (2007), 7313–14.

[Nemcova 2005] Nemcova, Petra and Scovell, Jane. *Love Always, Petra*, New York: Warner Books, 2005.

[Orell 2007] Orell, David. *Apollo's Arrow: The Science of Prediction and the Future of Everything*, Toronto: Harper Collins, 2007.

[Pielke 2005] Pielke, Jr, R. A., Landsea, A., Mayfield, M., Laver, J., and Pasch, R. Hurricanes and Global Warming, *Bulletin of the American Meteorological Society* (November 2005), 1571–5.

[Pilkey 2007] Pilkey, Orrin and Pilkey-Jarvis, Linda. *Useless Arithemetic: Why Environmental Scientists Can't Predict the Future*, New York: Columbia University Press, 2007.

[Poston 1978] Poston, T. and Stewart, I. *Catastrophe Theory and its Applications*, San Francisco: Pitman, 1976.

[Poundstone 2005] Poundstone, William. *Fortune's Formula*, New York: Hill and Wang, 2005.

[Ramanathan 2006] Ramanathan, Veerabhadran. Global Warming, *Bulletin of the American Academy* (Spring 2006), 36–8.

[Raspe 1895] Raspe, Rudolph Erich. *The Surprising Adventures of Baron Münchausen*, London: Cassel, Petter, Galpin & Co., 1895.

[Rayleigh 1876] Rayleigh, Lord. On Waves, *Proceedings of the Royal Society* 1 (1876), 251–71.

[Richter 1977] Richter, Charles. *Bulletin of the Seismological Society of America* 36 (1977), 1.

[Russell 1845] Russell, John Scott. Report on Waves, *Report of the Fourteenth Meeting of the British Association for the Advancement of Science, York, September 1844* (London 1845), 311–90, plates XLVII–LVII.

[Seager 2006] Seager, Richard. Climate Mythology: The Gulf Stream, European Climate and Abrupt Change, Lamont-Doherty Earth Observatory, Columbia University, <http://www.ldeo.columbia.edu/res/div/ocp/gs/>.

[Shiller 2000] Shiller, Robert. *Irrational Exuberance*, Princeton, NJ: Princeton University Press, 2000.

[Sigurdsson 1999] Sigurdsson, Haraldur. *Melting the Earth: The History of Ideas on Volcanic Eruptions*, Oxford, New York: Oxford University Press, 1999.

[Sornette 2003] Sornette, Didier. *Why Stock Markets Crash: Critical Events in Complex Financial Systems*, Princeton, NJ: Princeton University Press, 2003.

[Sparks 2001] Sparks, Stephen. Dynamic Pulsating Eruptions, *FirstScience.com*, 6 (January 2001).

[Steel 1995] Steel, Duncan. *Rogue Asteroids and Doomsday Comets: The Search for the Million Megaton Menace That Threatens Life on Earth*, New York: John Wiley & Sons, 1995.

[Stein 2003] Stein, Seth and Wysession, Michael. *An Introduction to Seismology, Earthquakes, and Earth Structure*, Oxford: Blackwell Publishing, 2003.

[Stokes 1880] Stokes, G. G. *Mathematical and Physical Papers*, Cambridge: Cambridge University, 1880.

[Sun 2006] Sun, Jun, Earl, David J., and Deem, Michael W. Localization and Glassy Dynamics in the Immune System, *Modern Physics Letters B*, 20 (2006), 63–95.

[Sutliff 1925] Sutliff, Tracy. Revival in All Industries Exceeds Most Sanguine Hopes, *New York Herald Tribune* (2 January 1925).

[Thom 1972] Thom, Réne. *Stabilité Structurelle et Morphogenese,* Reading: MA: W. A. Benjamin, 1972.

[Velikovsky 1950] Velikovsky, Immanuel. *Worlds in Collision,* New York: Doubleday, 1950.

[Velikovsky 1955] Velikovsky, Immanuel. *Earth in Upheaval,* New York: Doubleday, 1955.

[Verma 2005] Verma, Sundra. *The Tunguska Fireball: Solving One of the Greatest Mysteries of the 20th Century,* Toronto: Icon Books, 2005.

[Verschuur 1996] Verschuur, Gerrit L. *Impact! The Threat of Comets and Asteroids,* New York: Oxford University Press, 1996.

[Visher 1949] Visher, S. S. Bailey Willis, *Annual Association of American Geographers* 39/4 (1949), 291–2.

[Wang 2006] Wang, Kelin *et al.* Predicting the 1975 Haicheng Earthquake, *Bulletin of the Seismological Society of America* 96 (2006), 757–95.

[Weart 2003] Weart, Spencer. *The Discovery of Global Warming,* Cambridge, MA: Harvard University Press, 2004.

[Wente 2007] Wente, Margaret. A Questionable Truth, *The Globe and Mail* (27 January 2007), F1 and F7.

[Winchester 2003] Winchester, Simon. *Krakatoa: The Day the World Exploded—27 August 1883,* New York: Viking Penguin, 2003.

[Woo 1999] Woo, Gordon. *The Mathematics of Natural Catastrophes,* River Edge, NJ: World Scientific, 1999.

[Zeeman, 1976] Zeeman, Eric Christopher. Catastrophe Theory, *Scientific American* 234 (1976), 65–83.

INDEX